AS

Instant
Revision

Biology

Steve Potter

Series Editor: Jayne de Courcy

Published by HarperCollins*Publishers* Ltd
77-85 Fulham Palace Road
London W6 8JB

©HarperCollins*Publishers* 2002

First published 2002

ISBN 0 00 712424 4

British Library Cataloguing in Publication Data
A catalogue record for this book is available from the British Library.

Edited by Eva Fairnell
Production by Kathryn Botterill
Design by Gecko Ltd
Illustrations by Gecko Ltd
Cover design by Susi Martin-Taylor
Printed and bound by Printing Express Ltd, Hong Kong

You might also like to visit:
www.**fire**and**water**.com
The book lover's website

Steve Potter
Series Editor:
Jayne de Courcy

Instant Revision

AS Biology

Contents

Get the most out of your Instant Revision pocket book

1 **Maximize your revision time.** You can carry this book around with you anywhere. This means you can spend any spare moments revising.

2 **Learn and remember what you need to know.** The book contains all the really important facts you need to know for your exam. All the information is set out clearly and concisely, making it easy for you to revise.

3 **Find out what you don't know.** The *Check yourself* questions help you to discover quickly and easily the topics you're good at and those you're not so good at.

What's in this book

1 The content you need for *your* AS exam

- Important biological concepts are explained concisely to give a clear understanding of key processes.
- The table on page vi shows which chapters cover the modules in *your* specification, so only revise those topics you will be examined on.

2 Check yourself questions – find out how much you know and improve your grade

● The Check yourself questions occur at the end of each short topic.

● The questions are quick to answer. They are not actual exam questions, but the author is an examiner and has written them in such a way that they will highlight any vital gaps in your knowledge and understanding.

● The answers are given at the back of the book. When you have answered the questions, check your answers with those given. The examiner's hints give additional help with aspects of the answers which you might have had difficulty with.

● There are marks for each question. If you score very low marks for a particular Check yourself page, this shows that you are weak on that topic and need to put in more revision time.

Revise actively!

● **Concentrated, active revision** is much more successful than spending long periods reading through notes with half your mind on something else.

● The chapters in this book are quite short. For each of your revision sessions, choose a couple of topics and concentrate on reading and thinking them through for **20–30 minutes**. Then do the Check yourself questions. If you get a number of questions wrong, you will need to return to the topics at a later date. Some Biology topics are hard to grasp but, by coming back to them several times, your understanding will improve and you will become more confident about using them in the exam.

● Use this book to revise either on your own or with a friend!

The content you need for *your* AS exam

Speci-fication	Module	Instant Revision AS Biology chapter number																		
		1	2	3	4	5	6	7	8	9	10	11	12	13	14	15	16	17	18	19
AQA Spec.A	1†	x	x	x		x			x	x*										
	2				x		x	x								x*				
	3‡				x		x	x		x*						x*		x	x	x
AQA Spec.B	1	x	x	x		x			x	x*		x								
	2				x		x	x									x*			
	3a								x*	x*	x									
OCR	1	x	x	x		x	x			x*				x						
	2				x				x	x		x							x*	
	3								x*	x*	x							x*	x*	x
Edexcel	1†	x	x	x	x	x	x		x	x*		x								
	2B								x	x					x	x				
	2HB‡								x	x			x	x	x	x	x			
	3†												x	x						
WIEC	1	x	x	x	x	x	x	x	x	x	x	x*	x*							
	2					x	x			x				x						

† Modules which are common to Biology and Biology (Human) specifications.

‡ Modules which are part of Biology (Human) specifications only.

* Only part of this chapter is required in the module: the sections needed are indicated in the text.

Cells are the basic units of organisms. In many organisms, cells are organised into **tissues** and tissues into **organs**. Several organs often link into an **organ system**.

Most cells are microscopic, and to understand the images of cells that are produced by microscopes you need to understand a bit about how microscopes work.

Microscopes

Microscopes produce magnified images of specimens. Two main types are the **light microscope** and the **transmission electron microscope**. The table compares these two types of microscope.

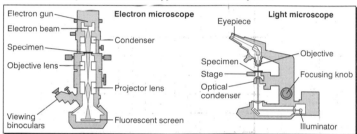

Feature	Electron microscope	Light microscope
Resolution (ability to separate points close together)	Much higher (the wavelength of electrons is much shorter than light)	Lower (light has a longer wavelength than electrons)
Magnification	Much higher, e.g. 500 000 times	Lower, e.g. 1500 times
Specimen	Only dead	Alive or dead
Preparation and appearance of specimen	Preparation involves staining with heavy metal salts and vacuum treatment, which can make the specimen's appearance unnatural	Preparation involves natural biological stains, so the specimen's appearance is likely to be natural
Size of specimen	Must be very thin: only a section of a cell	Thicker: can view a whole cell
How image is seen	Regions where electrons pass through appear light	Regions where light passes through appear light

MICROSCOPES AND CELLS (2)

A **scanning** electron microscope, like a transmission electron microscope, uses a beam of electrons, but the image is formed from **reflected** electrons. This reduces the resolution and magnification at which the microscope is effective. However, it can be used to look at detailed 3-D images of small objects, not just slices through a cell.

$$\text{Magnification of a microscope} = \frac{\text{apparent size}}{\text{actual size}}$$ (note: **both** sizes must be measured in the **same units**)

The units used could be nm or µm. 1 mm = 1000 µm. 1 µm = 1000 nm.

Cells

The two main types of cells are **prokaryotic** cells, e.g. bacterial cells, and **eukaryotic** cells, e.g. animal and plant cells. Prokaryotic cells are small, simple cells. Eukaryotic cells are larger and contain more complex cell organelles.

Cell organelle	Prokaryotic cells	Eukaryotic cells Plant cells	Animal cells
Cell wall	Present: not cellulose	Present: cellulose	Absent
Nucleus	Absent	Present	Present
DNA	Naked, a continuous loop in the cytoplasm	Bound to histones in chromosomes	Bound to histones in chromosomes
Chloroplast	Absent	Present	Absent
Mitochondrion	Absent	Present	Present
Endoplasmic reticulum (ER)	Absent	Present	Present
Golgi apparatus	Absent	Present	Present
Ribosomes	Small (70S)	Large (80S)	Large (80S)
Lysosomes	Absent	Present	Present

Viruses are not true cells because they have no organelles.

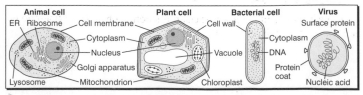

Animal cell — ER, Ribosome, Cell membrane, Cytoplasm, Nucleus, Golgi apparatus, Lysosome, Mitochondrion

Plant cell — Cell wall, Vacuole, Chloroplast

Bacterial cell — Cytoplasm, DNA, Protein coat

Virus — Surface protein, Nucleic acid

Organelle	Function
Nucleus	DNA in the nucleus directs protein synthesis through messenger RNA, and replicates in cell division. The nucleolus synthesises ribosomal RNA.
Ribosomes	Synthesise proteins from free amino acids brought by transfer RNA.
Mitochondria	Sites of many reactions of aerobic respiration (Krebs cycle in the matrix and electron transport on the cristae).
Chloroplasts	Sites of photosynthesis.
Endoplasmic reticulum (ER)	Transports synthesised proteins to the Golgi apparatus. Rough ER is associated with ribosomes.
Golgi apparatus	Modifies proteins and produces vesicles containing proteins to be released from the plasma membrane by exocytosis.
Plasma membrane	Controls entry and exit from cells by simple diffusion, facilitated diffusion, active transport, osmosis, endocytosis and exocytosis.
Microvilli	Increase the surface area of the cell, so increasing rate of absorption.
Lysosomes	Contain lytic (splitting) enzymes that digest worn-out tissue and foreign material in the cell.
Cell wall	Provides rigidity and support.
Cilia	Where present, they can move the whole cell (if it is not a fixed cell) or create currents outside the cell to move other objects.

Cell fractionation is a process that separates organelles, e.g. to create a pure sample of mitochondria. Cells are homogenised then centrifuged. More dense organelles settle out at lower speeds. The sequence of separation of organelles, from the slowest to highest speed, is: nucleus – chloroplasts – mitochondria – lysosomes – membranes – ribosomes. So the nucleus is more dense than ribosomes.

Specialised cells

Many cells are adapted for a particular function. Individual specialisations are dealt with in relevant chapters.

Tissue comprises a group of similar specialised cells carrying out the same function. For example:

- **smooth muscle** cells are all spindle shaped and contractile
- **palisade mesophyll** cells are elongated and contain many chloroplasts for photosynthesis
- **xylem** cells are tubular, empty cells for the transport of water in plants
- **blood** contains different types of cells, one type being for transport of substances around the body.

An **organ** is a structure with a clearly defined function made from several tissues.

A capillary is not an organ because it contains only one tissue, epithelium.

Organ	Function	Tissue	Contribution to overall function
Artery	Transports blood	Fibrous tissue	Provides support and holds artery open
		Smooth muscle	Allows constriction/dilation of artery
		Elastic tissue	Allows artery to return to smaller size
		Endothelium	Smoothness allows easy flow of blood
Leaf	Photosynthesis/ transpiration	Palisade mesophyll	Many chloroplasts for photosynthesis
		Spongy mesophyll	Air spaces for gas movement
		Epidermis	Restricts water loss; stomata help gas exchange
		Xylem	Transports water and mineral ions to leaf
		Phloem	Transports organic molecules to and from leaf

In an **organ system**, several organs are linked for a major biological function. The heart, arteries, capillaries and veins form the **circulatory system**. The mouth, oesophagus, stomach, intestines and associated glands form the **digestive system**.

Check yourself

1 (a) Explain why a scanning electron microscope produces more detailed images of cells than a light microscope. (2)
   (b) Give two disadvantages of using a transmission electron microscope. (2)

2 On an electron micrograph, the width of a cell is 6.5 cm. The actual width is 130 μm. Calculate the total magnification. Show your working. (3)

3 The drawing shows an electron micrograph of an animal cell.
   (a) Name the organelles labelled A, B and C. (3)
   (b) Give two pieces of evidence from the drawing that suggest that this is an active cell. (2)

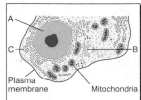

4 In cell fractionation, a suspension of cells in an ice-cold isotonic buffer is homogenised. The suspension is then centrifuged at increasing speeds.
   (a) Why is a buffer solution used? (2)
   (b) Why is the buffer solution ice-cold and isotonic with the cells? (4)

5 (a) Explain what is meant by the terms tissue and organ. (2)
   (b) Explain why an artery is called an organ while a capillary is not. (2)
   (c) Explain why blood is an unusual example of a tissue. (2)

6 The diagram shows a cross-section through a leaf.

   Explain how the structure of the leaf allows efficient photosynthesis. (6)

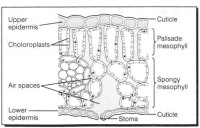

The answers are on page 104.

BIOLOGICAL MOLECULES (1)

Carbohydrates

Carbohydrates are sugars. They contain **C**, **H** and **O**, with H and O in the ratio 2:1. **Monosaccharides** are sugars with a single ring of atoms, e.g. ribose, which is a **pentose** (5-carbon) sugar ($C_5H_{10}O_5$), and glucose, which is a **hexose** (6-carbon) sugar ($C_6H_{12}O_6$).

α-Glucose β-Glucose

Disaccharides are sugars with two monosaccharide rings joined together. Sucrose and maltose have the formula $C_{12}H_{22}O_{11}$. As the monosaccharides join by **condensation**, a molecule of water is lost and a **glycosidic bond** is formed.

Maltose

α-Glucose α-Glucose
Glycosidic bond

Sucrose

Glucose   Fructose
Glycosidic bond

Polysaccharides are **polymers** of monosaccharides (usually glucose).

Starch

α-Glucose

6

Lipids

Lipids are triglycerides (fat and oils), phospholipids and other related compounds. They contain C, H and O, but have a higher ratio of H to O than carbohydrates.

Triglyceride molecules consist of three fatty acid molecules joined by **ester bonds** (formed by condensation) to a glycerol molecule. Fatty acids with double bonds in the hydrocarbon chain are unsaturated; others are saturated fatty acids.

Phospholipid molecules have two fatty acid molecules and a phosphate group joined to glycerol. The molecules are **polar** and form **bilayers** in aqueous systems.

Proteins

Proteins contain C, H, O and N (sometimes S). They are polymers of **amino acids** joined by **peptide bonds** formed by condensation.

Proteins have a **primary**, **secondary** and **tertiary** structure. The primary structure (the sequence of the amino acids) determines the secondary structure, which is usually an α-helix, held in place by hydrogen bonds; this then determines any further bonding and folding, the tertiary structure, held in place by ionic and disulphide bonds.

7

BIOLOGICAL MOLECULES (3)

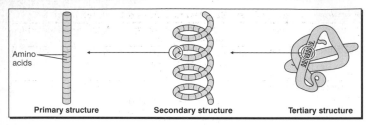

Primary structure	**Secondary structure**	**Tertiary structure**

Amino acids

Condensation joins molecules; **hydrolysis splits** molecules.

Glycosidic bond

Peptide bond

Ester bond

	Starch	Glycogen	Cellulose	Lipid	Protein
Polymer?	Yes, monomer is α-glucose	Yes, monomer is α-glucose	Yes, monomer is β-glucose	No	Yes, monomer is amino acid
Branching present?	Some	Much	None	None	None
Function	Storage carbohydrate in plants	Storage carbohydrate in animals	Fibrils form a meshwork in plant cell wall	Energy store; phospholipids in membranes	Enzymes, ion pores, carrier proteins
Adaptation to function	Compact, insoluble	Compact, insoluble	Fibrils form a strong mesh	Phospholipids are part polar	Specific 3-D shape

You need to know the tests shown opposite to identify biological molecules.

Test for	Name of test	How test is carried out	Result
Starch	Iodine	Add iodine solution to substance/solution	Yellow → blue/black
Reducing sugar, e.g. glucose	Benedict's test	Heat substance with Benedict's solution	Blue → yellow/orange/red precipitate
Non-reducing sugar, e.g. sucrose	Benedict's test	1 Test as above	No change
		2 Hydrolyse: boil with HCl (splits disaccharide molecule into monosaccharides)	
		3 Neutralise: add sodium hydrogen carbonate	
		4 Retest with Benedict's solution	Yellow/red precipitate
Protein	Biuret test	Add Biuret solution (or sodium hydroxide and 1% copper sulphate), wait for a minute	Blue → purple/mauve
Lipid	Emulsion test	1 Mix substance with ethanol (alcohol)	
		2 Filter contents into a test tube of water	Clear → milky emulsion

Chromatography is a process that separates the components of a mixture. Amino acids can be separated by 1-D or 2-D chromatography. The distance the solvent and each amino acid travels is measured. The ratio of

$$\frac{\text{distance amino acid travels}}{\text{distance solvent travels}}$$ is called the R_f value.

In 2-D chromatography, the process is repeated twice. Some amino acids have the same R_f value in one solvent but not in another. A square piece of paper is used; one solvent is used in the first run, then the paper is turned $90°$ and run in a second solvent.

Water

Water contains **H** and **O** as H_2O. It is the most common substance. **Hydrogen bonds** form between adjacent molecules. It is a good solvent for ions and molecules with –OH groups, e.g. glucose and amino acids.

Property	Biological significance
Liquid 0°C → 100°C	Does not change state at most normal temperatures
High specific heat capacity	Temperature changes only slowly; aquatic organisms are not subject to rapid fluctuations, which would affect enzyme-controlled reactions
Freezes at surface	Aquatic organisms are able to survive underneath the ice
Good solvent	Able to dissolve and transport many molecules
Forms hydrogen bonds	Strong cohesion between molecules, which is important in transpiration
High surface tension	Small organisms can live at the surface: they 'walk on water'
Metabolically active	Reactant in photosynthesis and all digestive reactions

Mineral ions

Mineral ions are needed in small quantities by most organisms.

Mineral	Use
Calcium, Ca^{2+}	Needed by plants for cell walls, and by vertebrate animals for bones, teeth, clotting of blood and muscular contraction
Magnesium, Mg^{2+}	Needed by plants to make chlorophyll, and by vertebrate animals for bones and teeth
Potassium, K^+	Needed by plants to control stomatal opening and activate over 40 different enzymes, and by animals for nerve impulses
Sodium, Na^+	Needed by some plants (e.g. maize and sorghum) for photosynthesis and osmotic control, and by animals for nerve impulses
Chloride, Cl^-	Needed by plants as an activator of photosynthesis and for osmotic control, and by animals for nerve impulses
Nitrate, NO_3^-	Needed by plants to make amino acids, but not needed by most animals
Phosphate, PO_4^{3-}	Needed by all organisms to make DNA, ATP and phospholipids for cell membranes.

1 (a) What are monosaccharides and disaccharides? (4)
   (b) How does the structure of glycogen differ from starch and cellulose? (3)
   (c) Name the bond that links glucose units in glycogen. (1)

2 (a) Give two ways in which starch is an ideal storage compound in plants. (2)
   (b) How could you test some onion cells to see if they contained starch? (2)
   (c) How are cellulose molecules arranged in the cell walls of plants? (3)

3 The diagram of an unsaturated fatty acid molecule is incomplete.

```
        H   H   H              H   H   H
        |   |   |              |   |   |
   H—C   C   C   C   C   C   C   C   C—
        |   |   |   |   |   |   |   |
        H   H   H   H   H   H   H   H
```

   (a) Copy and complete the diagram. (2)
   (b) How can saturated fatty acids be dangerous to human health? (2)
   (c) How many fatty acids are there in: (i) a triglyceride molecule; (ii) a phospholipid molecule? (2)

4 (a) Draw and name the molecules formed when two molecules of α-glucose are joined together. (4)
   (b) Name the process occurring. (1)
   (c) Name one reducing sugar and one non-reducing sugar. (2)

5 The diagram shows the result of 1-D chromatography of amino acids.
   (a) Explain two essential precautions when starting such an investigation. (4)
   (b) Calculate the R_f values for amino acids X and Y. (2)

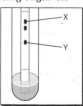

6 Explain three properties of water that are important to organisms. (6)

The answers are on page 104.

Enzymes are:
- **globular proteins**, with a specific tertiary structure (3-D shape) that has a precise area called an **active site**;
- **catalysts**, which speed up chemical reactions without being chemically altered;
- **specific** to a particular substrate because of the shape of their active site;
- **affected** by conditions such as temperature, pH, substrate concentration, enzyme concentration and the presence of inhibitors.

Enzymes as catalysts

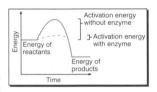

Enzymes are able to speed up chemical reactions by lowering the **activation energy** needed. Activation energy is the energy needed to make particles react when they collide. A fixed amount of energy in the environment can activate a certain number of reacting particles per second. If the activation energy is reduced to one-tenth, the same energy will activate 10 times as many particles: the reaction will be 10 times faster.

Specificity of enzymes

An enzyme's active site has a shape **complementary** to that of its substrate(s). Two ideas about how enzymes work are called **lock and key** and **induced fit**. Both share the ideas that the final shape of the active site and substrate are complementary and that enzyme and substrate bind to form an **enzyme–substrate complex**. Therefore the enzyme can only catalyse reactions between substrate particles with a shape that can bind to the active site. The enzyme is specific.

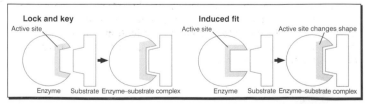

ENZYMES (2)

Factors affecting activity of enzymes

Temperature

Increasing temperatures up to an optimum increases enzyme activity because:
- the enzyme and substrate gain more **kinetic energy** and move faster;
- they collide more frequently as a result;
- more enzyme–substrate complexes are formed.

Increasing the temperature beyond the optimum causes a rapid decrease in enzyme activity because:
- the enzyme molecules vibrate more with the extra energy, distorting the shape of the active site by breaking the bonds that maintain the 3-D structure (it becomes **denatured**);
- the substrate molecule can no longer bind easily to the active site;
- fewer enzyme–substrate complexes are formed.

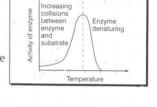

pH

A pH too far away from the optimum, which is often about pH 7, denatures the enzyme and reduces activity.

Enzyme concentration

Increasing the enzyme concentration makes more active sites available to bind with substrate, and so the rate of reaction will increase. This will continue until **all** substrate molecules bind **instantly** to the active sites: this rate cannot be exceeded, and a further increase in enzyme concentration will have no effect.

Substrate concentration

As the substrate concentration increases, the rate of reaction increases because:
● there are more collisions between enzyme and substrate;
● more enzyme–substrate complexes are formed.

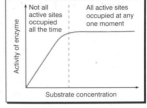

Above a certain substrate concentration, no further increase occurs because, at any time, all active sites are bound to substrate. No more substrate can bind until an active site becomes free. This represents the maximum reaction rate (**turnover rate**): the full capacity of the enzyme.

Inhibitors

Competitive (active site-directed) inhibitor molecules are similar in shape to substrate molecules and can bind to the active site of an enzyme. This prevents the substrate from binding and the reaction is inhibited. The extent of inhibition depends on the ratio of inhibitor and substrate molecules: 25% inhibitor and 75% substrate will mean that, at any one time, 25% of the active sites are bound to inhibitors and the reaction rate will be reduced by 25%.

Non-competitive (non-active site-directed) inhibitors bind to an **allosteric** site (a site away from the active site) of an enzyme, so changing the shape of the active site. The substrate can no longer bind to the enzyme.

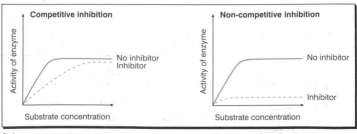

Enzymes in biotechnology

Enzymes are increasingly being used in biotechnological processes because:
- they catalyse reactions at moderate temperatures, saving heating costs;
- each is specific and catalyses only one reaction, so there are few by-products;
- they have a high turnover rate (catalyse thousands of reactions per second), so little is needed;
- they are often cheaper than conventional inorganic catalysts.

Biotechnologists often prefer to use isolated enzymes rather than whole micro-organisms because:
- some of the resources needed for the process are used by micro-organisms to produce their own biomass;
- micro-organisms produce many by-products, so purification is more costly;
- it is easier to optimise conditions for a single enzyme-controlled reaction than for all the reactions involved with a micro-organism.

```
        Small culture
        of bacteria
             │ Innoculate
             ▼
     Fermenter for bulk
     culture maintains
     ideal conditions of
     temperature, pH,
     O₂, nutrients.
     Bacteria multiply
     and secrete
     enzyme
      │              │ Filter
      ▼              ▼
  Bacteria       Enzyme
                 in solution
                    │ Purify
  Isolation of      ▼
  extracellular  Pure enzyme
  enzymes        in solution
                    │ Dry
                    ▼
                 Dry enzyme
                    │ Packaging
                    ▼
                 Packets of
                 dry enzyme
```

Some enzymes found in micro-organisms are **extracellular enzymes** (secreted by the micro-organisms); others are **intracellular** (remain in the micro-organisms). Isolating intracellular enzymes is more costly, because the cells have to be burst open and then undergo complex **downstream processing** (filtration purification) to recover the enzyme.

Immobilised enzymes are enzymes entrapped either in a medium (e.g. alginate beads and polymer microspheres) or on the surface of a

ENZYMES (5)

Enzyme	Source	Process	Basis of process
Pectinase	Bacteria	Extracting fruit juices	Digestion of pectin in cell walls
Protease	Bacteria	In washing powders to remove protein stains (e.g. egg yolk, blood)	Proteins digested to soluble amino acids
Lactase	Yeast	Making lactose reduced milk	Digested to glucose and galactose
Glucose isomerase	Bacteria	Production of fructose syrup as a sweetener	Glucose converted to fructose, which is sweeter so less can be used

matrix (e.g. collagen fibre matrix). Using immobilised enzymes has a number of advantages:

- the enzymes do not contaminate the product, so downstream processing is cheaper;
- continuous production is possible, and running costs are low as there are no repeated start-ups;
- the enzymes can be re-used many times, which is cheaper than periodic replacement;
- the enzymes are more stable when immobilised and can be used at higher temperatures, giving faster reaction rates.

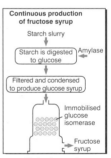

Continuous production of fructose syrup

Biosensors are biological devices used to monitor processes or reactions. Enzyme reactions can be used to produce a colour change or a voltage change in a biosensor. **Colour change** biosensors allow convenient, one-off, assessments of the level of a specific substance, e.g. **clinistix strips** contain two immobilised enzymes that are used to detect the level of glucose in a liquid such as urine. **Voltage change** biosensors can monitor levels of a substance continuously. A biosensor of this type using glucose oxidase can continuously monitor levels of glucose.

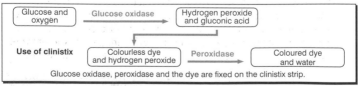

Use of clinistix

Glucose oxidase, peroxidase and the dye are fixed on the clinistix strip.

1 Explain how enzymes are able to catalyse (speed up) biological reactions. (2)

2 **(a)** What is meant by the 'specificity of enzymes'? (1)
  (b) How does the lock and key hypothesis account for the specificity of enzymes? (3)
  (c) How is the induced fit hypothesis different from the lock and key hypothesis? (2)

3 The graphs show the effects of temperature and substrate concentration on the activity of an enzyme.

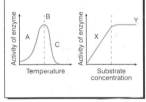

  (a) **(i)** Explain the effect of temperature over the region marked A. (3)
    (ii) Name the temperature marked B. Explain your choice. (2)
    (iii) What is happening to the enzyme at temperatures marked C? Why? (3)
  (b) **(i)** Explain the effect on enzyme activity of increasing the substrate concentration over the region marked X. (3)
    (ii) Why is there no change in the activity of the enzyme when the substrate concentration increases over the region marked Y? (2)

4 Complete the following paragraph.

Enzymes have a region called the .... which has a specific shape. Substrate molecules have a shape which is .... to this and so can bind with it to form an ....
.... Some inhibitor molecules have a shape similar to the substrate. These are called .... inhibitors. They can fit into the .... of the enzyme and block it. A higher proportion of inhibitor to substrate causes ....
inhibition. Other inhibitors bind to a different part of the enzyme called the .... This changes the .... of the

The answers are on page 105.

.... so that the substrate cannot bind. This inhibitor is called a .... inhibitor. This inhibition is .... of substrate concentration. (11)

5 The graph shows the effect of an inhibitor on the activity of an enzyme.

Does the graph illustrate competitive or non-competitive inhibition? Explain your answer. (3)

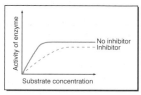

6 The flow chart shows the main stages in extracting an extracellular enzyme from a micro-organism.
   (a) Which of the processes A – E are involved in downstream processing? (1)
   (b) Why is it more costly to isolate an intracellular enzyme? (2)
   (c) Give two reasons for preferring to use isolated enzymes rather than whole micro-organisms in biotechnological processes. (2)

7 Many enzymes are immobilised for use in biotechnological processes.
   (a) Give three advantages of using immobilised enzymes. (3)
   (b) Give two different examples of the use of immobilised enzymes. (2)

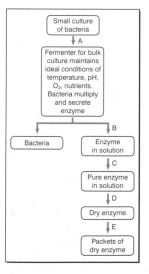

The answers are on page 106.

THE CELL CYCLE (1)

Chromosomes are structures found in the nucleus of a cell. They consist of one long DNA molecule wound around proteins called **histones**. This is shown very simplified in the diagram.

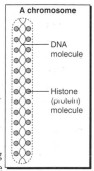

A chromosome

DNA molecule

Histone (protein) molecule

Genes are a section of DNA that occurs at a specific point, **locus**, along a chromosome and codes for a particular protein. **Alleles** are alternative forms of a gene.

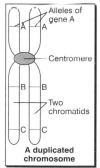

Alleles of gene A

Centromere

Two chromatids

A duplicated chromosome

A **homologous pair** of chromosomes is a pair of chromosomes that have genes controlling the same features along their length. They may not have the same alleles of those genes.

Chromatids are structures formed when a single chromosome replicates itself. The two chromatids are held together by a **centromere** to form a duplicated (or double) chromosome. These chromatids do have the same alleles of the genes.

A **diploid cell** has all chromosomes in homologous pairs: two of each type of chromosome. Most cells are diploid cells. A **haploid cell** has only one chromosome from each homologous pair: only half the number of a diploid cell. Sex cells are haploid.

The cell cycle

The **cell cycle** is the sequence of events that occurs as a cell grows, prepares for and finally undergoes cell division by mitosis.

G₁ phase	A newly formed cell is small and produces proteins and new organelles to increase in size
S phase	Cell enlargement continues; the DNA replicates; each chromosome becomes a pair of chromatids joined by a centromere
G₂ phase	Cell enlargement slows down; spindle proteins are synthesised
Mitosis	Pairs of chromatids split into single chromosomes and equal numbers move to opposite poles of the cell to form new nuclei
Cytokinesis	The cell divides into two new cells

THE CELL CYCLE (2)

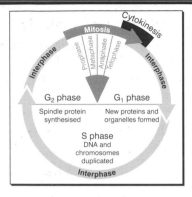

The graphs show changes in cell volume and DNA content during the cell cycle.

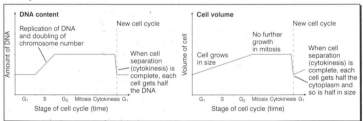

Some cells complete the cell cycle many times (divide repeatedly) to form cells for new growth or to replace damaged cells. Examples include:

● bone marrow cells (producing red and white blood cells);
● epithelial cells in the small intestine (replacing cells lost as food moves through);
● cells near the tips of plant roots (dividing to produce new growth).

These are relatively unspecialised cells. Specialised cells often only go through the cell cycle once, e.g. a nerve cell, once formed, cannot divide again.

Cell division – mitosis and meiosis

Both mitosis and meiosis involve a nuclear division (the chromosomes are divided between new nuclei) and cell division (new cells are formed).

Mitosis forms two diploid daughter cells, genetically identical to the parent cell.

Meiosis forms four haploid daughter cells that vary genetically.

Stage of process	Sequence for mitosis	Sequence for meiosis
Interphase	Chromosomes duplicate	Chromosomes duplicate
Prophase (I)	Nuclear membrane breaks down, chromosomes shorten and thicken They are made of two chromatids	Nuclear membrane breaks down, chromosomes shorten and thicken. They are made of two chromatids. Homologous chromosomes pair up
Metaphase (I) (= middle)	Chromosomes align on newly formed spindle	Chromosomes align on spindle: homologous chromosomes are still in pairs
Anaphase (I) (= apart)	Spindle fibres pull a chromatid from each chromosome to opposite poles	Spindle fibres separate homologous chromosomes
Telophase (I)	Two new nuclei form, each with the diploid number of single chromosomes	Two haploid nuclei form, chromosomes are double
Prophase II		Chromosomes shorten
Metaphase II		Chromosomes in each cell align on spindle
Anaphase II		One chromatid from each chromosome is pulled to opposite poles
Telophase II		Four haploid nuclei form, chromosomes are single
Cytokinesis	Two diploid cells formed	Four haploid cells formed

THE CELL CYCLE (4)

Mitosis, meiosis and life cycles

In **sexual reproduction**, two haploid **gametes** (e.g. a sperm and an egg with 23 chromosomes each) fuse to form a diploid **zygote** (with 46 chromosomes). The zygote divides by mitosis to form 2, 4, 8 ... and eventually billions of cells, which make up the body of the organism. All are diploid and genetically identical. When the organism is mature, cells in sex organs divide by meiosis to form haploid gametes and the cycle starts again.

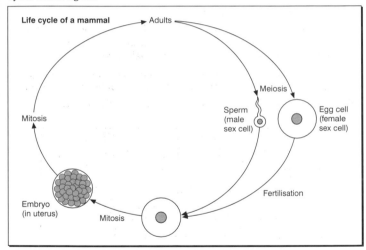

In a life cycle like the one above, meiosis, fertilisation and mitosis ensure that every cell in an organism (except the sex cells) is diploid, generation after generation. Each stage has a role:
● meiosis ensures that gametes are haploid;
● fertilisation results in a diploid zygote (the first cell of the new generation);
● mitosis ensures that all cells formed from this zygote are also diploid.

1 Label the diagram of a duplicated chromosome. (4)

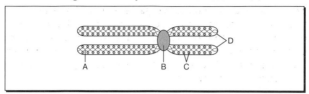

2 What are homologous chromosomes? (2)

3 Explain why the chromosomes that make up a homologous pair
may not have the same alleles of genes, whereas sister
chromatids always do have the same alleles. (4)

4 Name two cell organelles that will be particularly active during
the G_1 phase of the cell cycle. Give reasons for your choices. (4)

5 (a) Rearrange the drawings of the stages of mitosis into the
correct order. (1)

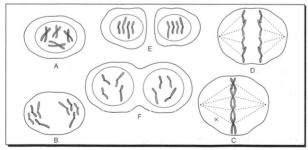

(b) Name the structure labelled X on diagram C. (1)

6 Give three differences between mitosis and meiosis. (3)

7 In the life cycle of a mammal, how do meiosis, mitosis and
fertilisation ensure that the chromosome number of ordinary
body cells is constant from generation to generation? (3)

The answers are on page 106.

A **plasma membrane** (or cell surface membrane) surrounds all cells. Cell organelles, e.g. nucleus, chloroplasts and mitochondria, also consist largely of membranes. The endoplasmic reticulum (ER) is an internal system of membranes.

Structure of plasma membranes

In the **fluid mosaic model of membrane** structure, phospholipids form a **bilayer** (double layer) with protein and other molecules bound to it in a continuous pattern, like a mosaic. The position of the protein molecules can change from time to time, so the pattern is 'fluid'. **Cholesterol** molecules reduce fluidity at higher temperatures and so help to maintain membrane structure.

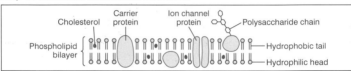

The phospholipid bilayer only allows small **non-polar** (non-charged) particles, e.g. water, and lipid-soluble particles, e.g. glycerol, to pass through freely.

Different proteins in the membrane have different functions:

- **glycoproteins** have a polysaccharide chain extending from the surface of the membrane and act as cell recognition sites;
- **receptor proteins** are proteins that have a specifically shaped binding site (**not** an active site) to bind with specific molecules, such as hormones;
- **ionic (hydrophilic) pore proteins** allow specific charged particles that could not pass through the phospholipid bilayer to pass through the channel they create, usually being selective on the basis of charge and size;
- **transport/carrier proteins** transport molecules across the membrane that are too large to pass by any other means, some carry particles down a concentration gradient (facilitated diffusion), others against the gradient (active transport).

Membranes serve many functions within a cell.

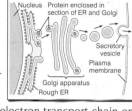

- They compartmentalise the cytoplasm, keeping metabolic processes separate, e.g. mitochondria contain all the chemicals needed for aerobic respiration.
- They form a surface on which enzymes can be bound, allowing sequential reactions to proceed effectively, e.g. the electron transport chain on the inner membrane of mitochondria.

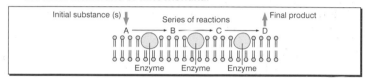

- They control the passage of substances into and out of a cell.

Methods of transport across a plasma membrane

Diffusion is a **passive** process, requiring no extra input of energy from respiration. There is a net movement of particles from a high concentration to a lower one due to random movement of the particles. Increasing the **kinetic energy** of the particles increases the diffusion rate as they move faster. When diffusion takes place across a boundary structure (e.g. plasma membrane or the epithelium of the alveolus), several factors affect the rate, including the:

- difference in concentrations either side of the boundary – the larger the difference, the faster the rate of diffusion;
- total surface area of the boundary – if there is more surface area, there are more places to cross and so diffusion will be faster;
- thickness of the boundary – a thicker boundary slows down diffusion.

The relationship between these factors is summarised in **Fick's Law**:

$$\text{Rate of diffusion} \propto \frac{\text{difference in concentration} \times \text{surface area of boundary}}{\text{thickness of boundary}}$$

Facilitated diffusion is also a passive process that depends only on the kinetic energy of the particles and takes place down a concentration gradient. Unlike simple diffusion, carrier proteins are needed to move the particles across the membrane. Fick's law can be modified for facilitated diffusion to read:

$$\text{Rate of diffusion} \propto \frac{\text{difference in concentration} \times \text{number of carrier proteins}}{\text{thickness of boundary}}$$

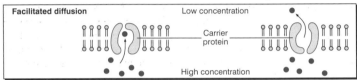

Active transport is not a passive process and requires an input of energy as ATP from respiration. Carrier proteins move particles against a concentration gradient.

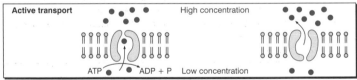

The diagram on the right shows the uptake of glucose by a cell lining the small intestine. Active transport pumps glucose into the blood stream. In doing so, it lowers the concentration of glucose in the cell, so more can enter by facilitated diffusion from the gut lumen. Facilitated diffusion would slow down and stop if the concentration in the cell rose to the level of that in the gut lumen.

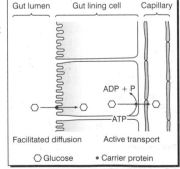

Osmosis is the movement of **water** through a **partially permeable membrane** from a high water potential (Ψ) to a lower water potential (effectively from a dilute solution to a more concentrated one). Ψ is measured in kiloPascals (kPa).

Pure water has a Ψ of 0 (zero). When solute particles dissolve in water, they attract water particles to form hydration shells round them and the water molecules don't move so freely. This reduces the Ψ, giving it a negative value. More solute makes the solution more concentrated and lowers ψ further, making it more negative.

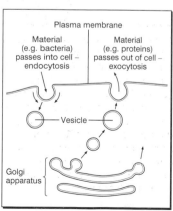

Cell A
$\psi = -11\,\text{kPa}$

Cell B
$\psi = -18\,\text{kPa}$

→ Water movement

If an animal cell gains too much water by osmosis, the increased volume may put too much pressure on the plasma membrane and cause it to burst. The cell walls of plant or bacterial cells help prevent this.

The processes of **endocytosis** and **exocytosis** move large particles into and out of a cell. In both, **vesicles** are formed that carry particles through the cytoplasm.

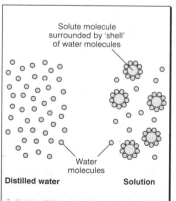

Solute molecule surrounded by 'shell' of water molecules

Water molecules

Distilled water   **Solution**

Plasma membrane

Material (e.g. bacteria) passes into cell – endocytosis

Material (e.g. proteins) passes out of cell – exocytosis

Vesicle

Golgi apparatus

Check yourself

1 **(a)** Name parts A – E on the diagram. (5)

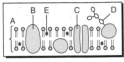

(b) Through which of the labelled parts might the following pass: a sodium ion Na$^+$, a glucose molecule, a glycerol molecule? (3)

(c) Why would a high temperature make the membrane freely permeable? (3)

2 **(a)** Give two functions of membranes in cells, other than transport. (2)

(b) Give four locations of membranes in a eukaryotic cell. (4)

(c) Which three of these locations are *not* common to prokaryotic cells? (3)

3 Diffusion rate $\propto \dfrac{\text{difference in concentration} \times \text{surface area of boundary}}{\text{thickness of boundary}}$ (Fick's law)

Why is diffusion efficient when gas exchange takes place in the alveoli? (3)

4 The diagram shows how amino acids pass from the gut lumen, through a cell in the small intestine and into the blood stream.

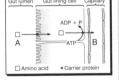

(a) Identify, with reasons, processes A and B. (4)

(b) Explain how a constant concentration gradient is maintained. (2)

5 The diagram shows three adjacent cells and their water potentials (Ψ).

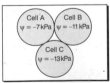

(a) Copy and draw arrows to show water movement between the cells. (3)

(b) Explain why red blood cells placed in distilled water burst. (3)

(c) Explain why red blood cells placed in strong salt solution shrivel. (3)

28

The answers are on page 107.

Nucleic acids are **polymers** of **nucleotides**, so can be called **polynucleotides**. Each nucleotide consists of three components: a pentose (5-carbon) sugar, a phosphate group and nitrogenous base.

DNA (Deoxyribo-Nucleic Acid)

DNA nucleotides contain the pentose sugar **deoxyribose**, a **phosphate**, and **one** of four nitrogenous bases: **adenine** (A), **thymine** (T), **cytosine** (C) or **guanine** (G). DNA molecules have two strands of nucleotides linked by hydrogen bonds.

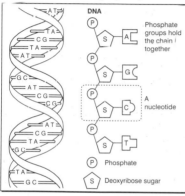

RNA (Ribo-Nucleic Acid)

There are three types of RNA: **messenger RNA** (mRNA), **transfer RNA** (tRNA) and **ribosomal RNA** (rRNA). All differ from DNA in several ways:

● RNA molecules are smaller;
● RNA molecules are single stranded;
● the base thymine is replaced by the base **uracil** (U);
● the pentose sugar is **ribose**.

DNA replication

If a cell is going to divide, the DNA must be copied. This is **DNA replication**. During interphase (see page 20) DNA undergoes **semi-conservative replication**:

1 the enzyme DNA helicase breaks hydrogen bonds and separates the DNA strands;

2 DNA polymerase then builds a new strand alongside each separated strand, and base pairing of A – T and C – G ensures that the new strand is complementary;

3 each new molecule of DNA contains one strand from the original DNA molecule and a newly synthesised complementary strand: they are identical to each other and to the original strand.

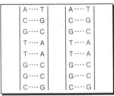

Meselson and Stahl's experiment provided evidence for semi-conservative replication. They grew bacteria in media containing either N^{14} or N^{15}, which became part of the bases in the DNA. N^{14} gave 'normal' or 'light' DNA, but N^{15} gave 'heavy' DNA. They were able to extract the DNA and centrifuge it in a special medium. Normal and heavy DNA settle to different positions.

Stage in experiment	Result from centrifuging	Explanation
1. Bacteria grown on N^{15} medium	'heavy' DNA	All the DNA produced is 'heavy' DNA
2. Bacteria transferred to N^{14} medium and reproduce once (1st generation)	'intermediate' DNA	DNA has replicated once; each original strand had N^{15}, each new strand has N^{14}. The DNA is intermediate in mass
3. Bacteria still in N^{14} medium and reproduce again (2nd generation)	'light' DNA 'intermediate' DNA	DNA replicates again. Each new strand is N^{14}, light DNA. Some DNA is 'all N^{14}', light DNA. Some is half N^{14}, half N^{15} DNA (intermediate in mass)
4. The bacteria continue to reproduce and the DNA replicates. All the new DNA strands are 'normal' (light). So nearly all of the DNA is totally normal, but a smaller and smaller percentage is intermediate DNA		

Protein synthesis

Proteins are polymers of amino acids that are synthesised in ribosomes. The **genetic code** for a protein molecule is carried in a gene in a DNA molecule. **Transcription** produces mRNA (a mobile copy of a gene) that carries the code to the ribosomes. tRNA brings the amino acids to a specific place on the ribosomes and the code is **translated** into a protein molecule.

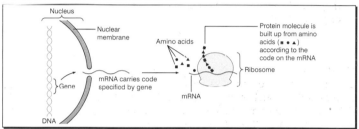

The **gene** (section of DNA) specifies the order in which specific amino acids must be assembled to make a particular protein. The code in the gene is determined by the sequence of bases in a DNA strand. The code is:

- **triplet** – a sequence of three bases codes for an amino acid;
- **degenerate** – there are more triplets than there are amino acids, so an amino acid can have more than one triplet, and some triplets do not code for any amino acid;
- **non-overlapping** – bases in one triplet do not also belong to another;
- **universal** – each triplet specifies the same amino acid in all organisms.

Transcription

1. A gene is activated: the two strands in this region of the DNA molecule part. Only the **coding strand** is transcribed.

2. **RNA polymerase** manufactures a single complementary strand by base pairing using RNA nucleotides: this is mRNA.

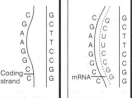

3 The mRNA molecule leaves the DNA and the nucleus
through pores in the nuclear membrane. Triplets of
bases in the mRNA called **codons** now carry the
genetic code for the protein.

C	G	G
G	C	C
A	T	U
A	T	C
G	C	U
G	C	C
C	G	G
C	G	G

Translation

One end of each tRNA molecule has an attachment site for a specific
amino acid, e.g. alanine. At the other end of the molecule is a triplet of
bases called an **anticodon** which is complementary to a codon on the
mRNA and can bind to it.

As the mRNA passes through a ribosome:

1 an anticodon on the tRNA binds to its codon and the amino acid is
held in place alongside the previous amino acid;
2 a peptide bond forms and links the two amino acids;
3 the tRNA is released and returns to the cytoplasm;
4 steps 1 – 3 repeat until the protein chain is assembled.

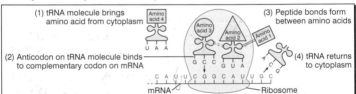

(1) tRNA molecule brings amino acid from cytoplasm

(2) Anticodon on tRNA molecule binds to complementary codon on mRNA

(3) Peptide bonds form between amino acids

(4) tRNA returns to cytoplasm

Point mutation

Point mutation	Explanation
Addition	An extra base is included in the DNA strand, altering the whole base sequence. This is called a **frameshift**. The code is completely changed.
Deletion	A base is omitted, again producing a frameshift.
Substitution	A different base is included from the original (e.g. T is included instead of A). Only one triplet is affected. Because the code is degenerate, substitutions may not alter the protein produced.
Inversion	The order of two bases in a triplet is reversed. Again, there is no frameshift and the protein specified may not be altered.

1 The diagram shows part of a DNA molecule.
   (a) Name the parts labelled A – F. (6)
   (b) Explain what is meant by 'complementary strands'. (3)
   (c) A sample of DNA contains 22% thymine. Calculate: **(i)** the % adenine; **(ii)** the % guanine. Show your working. (4)

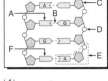

2 (a) Name A – E on the flow chart of protein synthesis. (5)
   (b) Where do transcription and translation take place? (2)

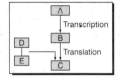

3 A section of mRNA has the base sequence CCG AAC GGA AUA UAC.
   (a) What is the base sequence on the DNA from which is was transcribed? (1)
   (b) What anticodon on tRNA will bind to the first and fourth codons? (2)

4 Complete the following paragraph.

DNA replication occurs during the .... phase of the cell cycle. It is called .... replication as each new DNA molecule contains one .... from the original molecule and one newly synthesised strand. The enzyme DNA .... breaks .... bonds to part the strands. DNA .... then makes a new .... strand for each using DNA .... It positions these according to the .... rule. This always pairs A with T and C with G. (9)

5 Look at the diagram of transcription. Name A, B, C and the bond forming at D. (4)

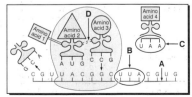

The answers are on page 108.

[AQA A, AQA B, OCR and WJEC only]

Genetic engineering describes the range of techniques used to manipulate DNA. It includes the transfer of genes from one species to another, to produce **transgenic organisms**. For example, bacteria can be engineered to produce a useful product.

Term	Meaning
Gene	A section of DNA coding for a specific protein
Restriction endonuclease	An enzyme that cuts DNA at a specific base sequence, e.g. GAATTC. Each time the sequence occurs, it cuts the DNA. If it occurs 5 times, there will be 5 cuts making 6 DNA fragments
Ligase	An enzyme that joins DNA sections
Vector	Something that transfers a gene from one cell to another, usually a bacterial plasmid or a virus
Plasmid	A small circular piece of DNA found in bacteria; it is not part of the main bacterial DNA
Sticky ends	Non-parallel ends of a DNA molecule formed when a restriction enzyme cuts the DNA. They allow other DNA to bond more easily than non-sticky ends

The main stages in the process are shown in the flow chart.

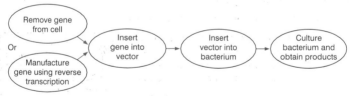

Removing a gene

The **donor cells** (cells containing the gene that is useful) are incubated with a **restriction endonuclease**. The enzyme cuts the DNA to leave sticky ends.

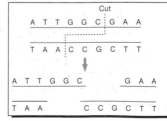

GENETIC ENGINEERING (2)

Making a gene

Think about the name of the enzyme: **reverse transcriptase**.
Transcription means making a mRNA copy of a section of DNA (a gene). Reverse it and you can make a DNA copy from the mRNA.

1 The mRNA coding for the appropriate protein is incubated with a reverse transcriptase and DNA nucleotides.

2 Reverse transcriptase creates a single complementary strand of DNA according to the base-pairing rule.

3 The reverse transcriptase is 'washed' out.

```
U  A  U  C  C  G  A  C    mRNA
            |  Reverse transcription
            |  + DNA nucleotides
U  A  U  C  C  G  A  C

A  T  A  G  G  C  T  G
```

4 The single-stranded DNA is incubated with DNA polymerase (see page 30) and DNA nucleotides.

5 DNA polymerase makes a second complementary strand that hydrogen bonds to the original strand.

```
A  T  A  G  G  C  T  G
            |  DNA polymerase
            |  + DNA nucleotides
A  T  A  G  G  C  T  G

T  A  T  C  C  G  A  C
```

Transferring a gene

1 Plasmids are obtained from bacterial cells.

2 To cut open the plasmid DNA, the same restriction endonuclease is used that was used to cut out the gene from the donor cell. This will leave the same sticky ends.

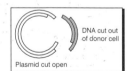

DNA cut out of donor cell

Plasmid cut open

3 The plasmids are incubated with the DNA (genes) obtained from the donor cell.

4 The DNA fragments from the donor cell bond to the plasmids.

5 The plasmids are incubated with bacteria. Some bacteria take up the plasmids. When they reproduce, their DNA replicates and so do the plasmids. All bacteria will contain the plasmids and the gene. The gene has been cloned.

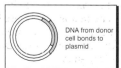

DNA from donor cell bonds to plasmid

Marker genes

A 'marker' gene is added to a plasmid at the same time as the desired gene, to enable a check that the bacteria have taken up the plasmid. The marker gene often gives resistance to an antibiotic. After incubation with the plasmid, the bacteria are cultured on a medium containing this antibiotic. Only those that have taken up the plasmid will be able to survive, as these contain the antibiotic resistance gene as well as the desired gene. The surviving bacteria can then be cultured on a large scale.

Culturing the bacteria

Bacteria are usually cultured in large, computer-controlled **fermenters**. The conditions are constantly monitored and adjusted, initially to provide optimum conditions for bacterial reproduction: during this phase little of the desired product is made. Then the key enzyme is 'switched on' and the conditions are re-optimised for maximum synthesis of the product.

Polymerase chain reaction (PCR) [AQA A and AQA B only]

This is a technique in which a sample of DNA is replicated without the need to clone it in a living organism. It is carried out in a PCR machine.

1   The PCR machine is loaded with: the DNA to be replicated; DNA nucleotides (the 'building blocks' for the new DNA); DNA primers (short sections of DNA to initiate replication); a **thermostable** DNA polymerase that allows continuous production despite the changes in temperature.

2   The PCR is heated to 95°C to separate the strands of DNA.

3   It is then cooled to 37°C to allow the primers to bind to the separated strands.

4   It is then heated to 72°C, the optimum for this DNA polymerase.

5   DNA polymerase uses the nucleotides to make complimentary strands.

6   The cycle is repeated.

In theory, each cycle doubles the amount of DNA as the molecules replicate. Over c cycles, the total amount of DNA produced (T) from an initial number (N) of molecules is given by the formula: $T = N \times 2^c$

Genetic fingerprinting

1 A sample of DNA is cut into fragments using a restriction endonuclease (if there is insufficient DNA more can be made using PCR).

2 The fragments are separated by gel electrophoresis.

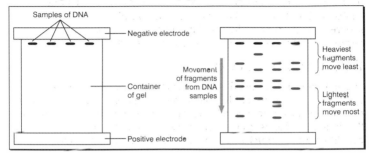

3 The fragments are then 'blotted' from the gel onto a membrane.

4 A radioactive DNA probe is applied to the membrane.

5 The probe binds to complementary regions (shown on X-ray film).

6 The genetic fingerprint is unique and can be used to identify a sample of DNA.

Gene therapy [WJEC and AQA B only]

This technique aims to treat genetic diseases, by inserting genes that will counter the effects of the genes causing disease. Cystic fibrosis (CF) is treated this way:

1 A healthy gene is cut from donor DNA using a restriction endonuclease.

2 The gene is transferred to a liposome (tiny fat globule) or a virus.

3 These are introduced into the patient by a nasal spray. They fuse with the membrane of epithelial cells, and the gene enters the cells.

4 The gene codes for the normal transport protein (not made in sufferers). This results in the formation of normal (non-viscous) mucus instead of the viscous mucus made in CF sufferers.

Check yourself

1 The flow chart shows the main stages in genetic production of a transgenic bacterium.
   (a) Name enzymes A and B. (2)
   (b) Why is it important that the DNA fragments and the open plasmids have sticky ends? (1)
   (c) What does 'recombinant plasmids' mean? (2)
   (d) Why can only recombinant bacteria grow on the antibiotic medium? (2)

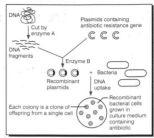

2 Complete the following paragraph.
   Reverse transcriptase is an .... that makes a single stranded .... copy of an mRNA molecule. To make a double stranded molecule, this must then be incubated with the enzyme .... as well as with .... The second strand is .... to the original single strand and the two are held together by .... bonds. (6)

3 Explain the benefits of dividing the culture of bacteria in a fermenter into a multiplication phase and one of production. (3)

4 (a) In the polymerase chain reaction, explain the benefit of using a thermostable DNA polymerase and DNA primers. (5)
   (b) What is the use of the PCR in genetic fingerprinting? (2)
   (c) How much DNA would be produced from 8 molecules in 6 cycles of the PCR? Show your working. (2)

5 (a) A restriction enzyme recognising the sequence CCGTTA, cuts a DNA sample into 8 fragments. How many times did the sequence occur? (1)
   (b) How does gel electrophoresis separate DNA fragments? (2)

6 In the treatment of cystic fibrosis by gene therapy, explain:
   (a) How the healthy gene could be removed from a cell. (2)
   (b) Why a virus is a suitable vector in this instance. (2)
   (c) Why this treatment can never give a permanent cure. (4)

The answers are on page 108.

The human breathing system

Ventilation moves air into and out of the alveoli where **gas exchange** takes place.

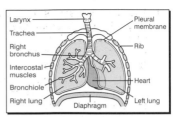

	Inspiration (inhaling)	Expiration (exhaling)
Intercostal muscles	Contract and raise ribs	Relax and allow ribs to fall
Muscle of diaphragm	Contracts, pulls diaphragm down	Relaxes, allows diaphragm to rise
Volume of thorax	Increases	Decreases
Pressure in thorax	Decreases	Increases
Air movement	In (from higher pressure outside)	Out (from higher pressure inside)

Breathing volumes	Definition
Tidal volume	The amount of air breathed in/out
Inspiratory reserve	The extra amount that can be inhaled besides the tidal air
Inspiratory capacity	Tidal volume + inspiratory reserve
Expiratory reserve	The extra amount that can be exhaled besides the tidal air
Vital capacity	Maximum possible air movement = inspiratory capacity + expiratory reserve
Residual air	Air that cannot be exhaled
Lung volume	Vital capacity + residual air

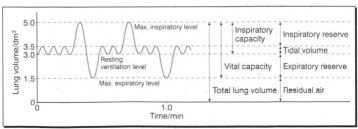

GAS EXCHANGE (2)

The amount of air breathed in/out per minute is the **pulmonary ventilation**. Pulmonary ventilation = **tidal volume** (volume per breath) × **rate of breathing**

Gas exchange in the alveoli

In the **alveoli**, oxygen and carbon dioxide **diffuse** in opposite directions down concentration gradients. These gradients are maintained by:
- circulation of blood – as blood around an alveolus becomes more oxygenated, it is moved on and replaced by deoxygenated blood;
- ventilation – as the air in the alveolus loses oxygen (and gains CO_2), it is replaced with air high in oxygen and low in CO_2.

Gas exchange in the alveoli is extremely efficient because:
- the alveoli collectively have a very large surface area
- each alveolus is surrounded by many capillaries

which means that there is extensive contact between capillaries and alveoli

- concentration gradients are maintained

- the alveolar epithelium is very thin (0.1 μm);
- the capillaries are very close to the alveoli;

which means that there is a short diffusion pathway

- the alveoli are moist (this allows carbon dioxide to diffuse more easily).

Control of breathing by the medulla of the brain

1 The **inspiratory** centre sends impulses that cause the intercostal muscles and diaphragm to contract. **Inspiration** starts, stimulating stretch receptors in the thorax wall.

2 Stretch receptors send **inhibitory** impulses to the inspiratory centre, reducing the number of impulses to the intercostal muscles and diaphragm.

3 Inhibition of the inspiratory centre increases as inspiration gets deeper. Eventually, no impulses are sent to the intercostal muscles and diaphragm and further inspiration is prevented.

4 **Expiration** occurs passively. Stretch receptors send fewer and fewer inhibitory impulses: at the end of expiration, inhibition of the inspiratory centre is removed.

In addition to the mechanism above, receptors in the aorta, carotid arteries and medulla can detect a rise in the amount of CO_2, e.g. during exercise. They send impulses to the respiratory centre to increase the rate and depth of breathing.

Transport of respiratory gases

Carbon dioxide is carried in the blood in three main ways:
- as CO_2 in solution in the plasma (5%);
- bound to haemoglobin in red blood cells and plasma proteins (10%);
- as hydrogencarbonate ions (HCO_3) in solution in the plasma (85%).

Oxygen is carried bound to **haemoglobin** in the red blood cells. Each haemoglobin (Hb) molecule has a **quaternary** structure of 4 polypeptide chains, each with a haem group to which oxygen can bind. Hb binds strongly to oxygen if there is a high oxygen tension in the surroundings, but only weakly if the oxygen tension is low. Hb can therefore 'load' oxygen in the lungs and 'unload' it in the tissues.

Polypeptide chains

Haem groups

The dissociation curve shows what % of Hb is bound to oxygen (the % saturation) under different oxygen concentrations. Most (95%) of the Hb binds to oxygen in the lungs, whereas in active tissues it can be as low as 20%. This means that 75% of the oxygen is unloaded to the tissues.

Dissociation curve

Saturation of Hb (%)

Oxygen tension (pO_2)/ kilopascals(kPa)

Tissues   Lungs

Exercise increases the respiration rate and the amount of CO_2 in the plasma. The plasma becomes more acidic, Hb binds more weakly to O_2 and releases more O_2 to the tissues: this is called the **Bohr effect**.

Fetal haemoglobin can bind more strongly to O_2 at low oxygen tensions than adult Hb. It is therefore able to load O_2 as the adult Hb unloads it at the placenta.

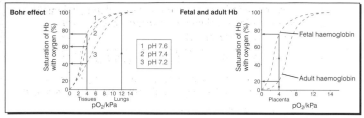

Bohr effect

1 pH 7.6
2 pH 7.4
3 pH 7.2

Saturation of Hb with oxygen (%)

Tissues   Lungs
pO_2/kPa

Fetal and adult Hb

Fetal haemoglobin

Adult haemoglobin

Placenta
pO_2/kPa

Gas exchange in plants

Most gas exchange in plants takes place through **stomata** (pores) in the leaf epidermis. The **cuticle** reduces evaporation of water from the epidermis. **Guard cells** control the width of stomata, and so control gas exchange. Most photosynthesis occurs in the **palisade mesophyll**. These cells have most chloroplasts. **Air spaces** in the **spongy mesophyll** allow movement of gases and water vapour through the leaf. **Xylem** in the vein (**vascular bundle**) transports water to the leaf. **Phloem** carries organic molecules, e.g. sugars and amino acids, to and from the leaf.

The main function of the leaf is **photosynthesis**. For this the plant needs CO_2 and H_2O as 'raw materials'. H_2O is always supplied by the xylem. CO_2 comes from respiration and from outside the leaf. Different gas exchanges occurr during the day and night.

Cuticle — Upper epidermis
Choloroplasts — Palisade mesophyll
Xylem —
Phloem — Spongy mesophyll
Cuticle — Lower epidermis
Stoma (plural stomata) allows exchange of CO_2 and O_2 and loss of water vapour
Air spaces allow rapid diffusion of CO_2 and O_2
Movement of water

The potassium ion (K^+) pump hypothesis explains how guard cells regulate the aperture (width) of the stomatal pore. During daylight:

- K^+ ions are actively pumped into guard cells;
- these accumulate and decrease the water potential of the cells;
- water enters and the guard cells swell;
- because of their unevenly thickened walls, they open the stomata. This is reversed during darkness.

Check yourself

1 Which three of the following make up the total lung volume? Vital capacity, residual volume, inspiratory capacity, tidal volume, expiratory reserve. (1)

2 Fick's law states diffusion rate $\propto \dfrac{\text{surface area} \times \text{concentration difference}}{\text{thickness of membrane}}$

Use Fick's law to explain how the structure and arrangement of alveoli and capillaries in the lungs enable maximum diffusion rate. (5)

3 Complete the following paragraph.

The basic breathing rate is controlled by the respiratory centre in the .... During inhalation, stretch receptors send .... to the .... These .... the inspiratory centre which sends fewer impulses to the .... and .... As the inhibition of the .... increases, inhalation slows and eventually stops. Exhalation occurs .... and reduces the number of impulses from the .... This means there is no .... of the inspiratory centre and .... begins again. (11)

4 (a) Explain how fetal Hb can load oxygen in the placenta. (3)
   (b) Why is iron essential in our diet? (1)

5 The diagram shows the distribution of K^+ ions in and around the guard cells of open and closed stomata. Explain how the differences in concentration bring about the opening and closing of the stomata. (8)

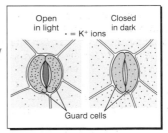

Guard cells

The answers are on page 109.

Organisms exchange materials with their environment. **Unicells** exchange materials, e.g. oxygen, through their plasma membrane. The **volume** of a unicell is a factor in determining how much oxygen it will need (the **demand rate**). The **surface area** determines how much it can get (the **supply rate**). The **surface area/volume ratio (SA/V ratio)** is a measure of whether supply is meeting demand.

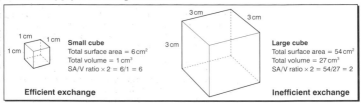

Small cube
Total surface area = 6 cm²
Total volume = 1 cm³
SA/V ratio × 2 = 6/1 = 6

Efficient exchange

Large cube
Total surface area = 54 cm²
Total volume = 27 cm³
SA/V ratio × 2 = 54/27 = 2

Inefficient exchange

Larger organisms are less able to exchange materials efficiently through their body surface. To solve this problem they have evolved either:
● a shape with a high SA/VR, e.g. are flat and thin, or
● specialised gas exchange organs with a high SA/V ratio together with a transport system.

The human circulatory system

You need to be able to name all the labelled parts.

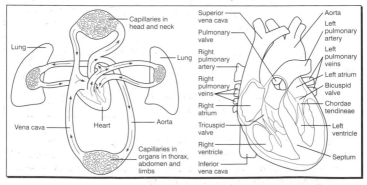

In the **cardiac cycle** a sequence of events pumps blood through the heart.

Stage of cycle	Events
Atrial systole (ventricles still in diastole)	Atria contract: blood forced through open atrio-ventricular valves. Ventricles relax as they fill with blood. Aortic and pulmonary valves closed
Ventricular systole (atrial diastole)	Atria relax: blood neither enters nor leaves. Ventricles contract: (a) pressure of blood in ventricles increases until greater than the pressure in atria, atrio-ventricular valves are forced shut; (b) when pressure exceeds that in major arteries, aortic and pulmonary valves are forced open, blood is ejected into arteries
Ventricular diastole (atria still in diastole)	Ventricles relax, walls exert no force. Atria still relaxed, walls exert no force: (a) blood enters atria but cannot pass into ventricles as atrio-ventricular valves are still shut; (b) pressure in atria builds and forces open valves, blood enters ventricles by passive (no contraction) ventricular filling.

Systole = contraction; diastole = relaxation of chamber.

Controlling the heartbeat

The heartbeat is **myogenic** (originates in the heart, not the nervous system):

- the **S–A node** in the right atrium produces an electrical impulse;
- the impulse passes through the atria along **Purkyne fibres**, stimulating the walls to contract;
- it can only pass to the ventricles through the **A–V node**, which conducts slowly and delays it;
- the impulse passes along **Bundles of His** (Purkyne tissue) and causes the ventricles to contract.

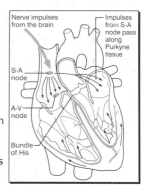

The delay at the A–V node allows the atria to empty before the ventricles contract. Although the heartbeat is myogenic, the autonomic nervous system influences the **heart rate** by causing the S–A node to generate more or less impulses per minute.

Sympathetic nerve impulses (cardiac nerve) from the cardiac centre in the medulla increase the heart rate. **Parasympathetic** impulses (vagus nerve) decrease the rate. The hormone **adrenaline** binds to β receptors in the S–A node and increases the rate. The **stroke volume** (the volume of blood pumped by **each** ventricle) is increased mainly by adrenaline and more blood returning to the heart (**venous return**).

Cardiac output is the amount of blood pumped per ventricle per minute:

Cardiac output = stroke volume × heart rate.

Blood vessels

Feature	Arteries	Arterioles	Capillaries	Veins
Structure of wall	Thick wall, much elastic tissue and smooth muscle, endothelium	Thinner, less elastic tissue, more muscle, endothelium	Endothelium only, with tiny gaps between cells	Thin wall, little muscle and elastic tissue, endothelium
Size of lumen	Large vessels with small lumen	Small vessels with small lumen	Microscopic with small lumen	Large vessels with large lumen
Direction of flow	Away from heart, to an organ	To capillaries within an organ	Around the cells within an organ	Away from organ, towards heart
Blood pressure	High but variable (pulsatile)	Lower than artery, less pulsatile	Falls through capillaries	Low: non-pulsatile
Adaptation to function	Muscle and elastic tissue, walls can stretch and recoil with high pressure. Even out pressure and act as secondary pumps	More muscle tissue: can dilate or constrict to alter blood flow, can redistribute blood	Extensive contact with cells (due to small size), 'leaky' walls allow tissue fluid to leave but large proteins are retained in plasma	Large lumen and thin walls offer least resistance to flow of low pressure blood. Valves prevent backflow

The effects of exercise

During exercise, muscles must receive an increased supply of oxygen and glucose and lose more carbon dioxide and heat. To achieve this:

- we breathe deeper and faster to increase the rate of gas exchange;
- the heart rate and stroke volume increase;
- blood is redistributed by constriction of arterioles to gut and skin and dilation of arterioles to muscles.

The **amount** of blood flowing to the brain and kidney remains unchanged; the **proportion** of the increased cardiac output **decreases**.

The composition of blood

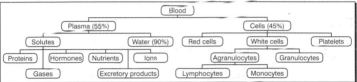

Cell	Features
Red blood cell	Enucleate, biconcave discs. No nucleus to allow more haemoglobin to be packed in for maximum oxygen transport. Shape gives large surface area and short diffusion distances to aid gas exchange
Lymphocytes	Cells about the same size as red cells, large round nucleus. B lymphocytes produce antibodies against specific antigens. T lymphocytes destroy virus-infected cells
Monocytes	Cells much larger than red blood cells, with large, kidney-shaped nucleus. Can migrate from blood into tissue to engulf and destroy bacteria, cancer cells and cell debris by phagocytosis
Granulocytes	Cells, e.g neutrophils intermediate in size between red cells and monocytes, with granular cytoplasm. Act as phagocytes or secrete histamines (anti-inflammatory substances)

Blood and defence against disease

[AQA A Biology (Human), Edexcel and OCR only]

Phagocytosis

Monocytes (enlarged to become macrophages) and granulocytes engulf foreign cells and release lytic enzymes to digest them. This is a non-specific response.

Humoral immune response

Antigens on the surface of micro-organisms are detected by
B lymphocytes. The B lymphocytes that make appropriate antibodies
are activated/sensitised and **clone** themselves repeatedly by mitosis:
this is **clonal selection**. Cloned B lymphocytes become either:
- **plasma cells**, which secrete antibodies that bind to the antigen and
  destroy the micro-organism, or
- **memory cells**, which remain in the circulation and mount a speedy
  secondary response to re-infection.

Cell-mediated immune response

Antigens on the surface of a micro-organism are detected by
T lymphocytes. Appropriate T lymphocytes are activated /sensitised
and clone themselves. The T lymphocytes multiply and become either:
- killer T lymphocytes, which secrete chemicals and kill virus-infected cells;
- helper T lymphocytes, which stimulate B cells and killer T cells to multiply;
- memory T cells, which remain in lymphoid tissue and stimulate a
  speedy secondary cellular response to re-infection.

Inflammation

Damaged tissue releases a number of chemicals, e.g. **histamine**, which:
- dilate arterioles, allowing more blood to the area, bringing more
  lymphocytes;
- make capillaries more permeable, allowing many proteins to
  escape, including fibrin which causes clotting, preventing bacteria
  from entering the damaged blood vessels and spreading to other
  parts of the body.

Active and passive immunity

Active immunity involves an organism making antibodies to an antigen.
It can be **natural**, e.g. the antigens on invading micro-organisms, or
artificial, e.g. exposure to antigens in a **vaccine. Passive immunity**
involves acquiring antibodies from another individual, e.g. across the
placenta, not made in an immune response.

Vaccination

Vaccines contain pathogens treated to stimulate an immune response to a particular micro-organism without causing the disease. They stimulate an immune response by specific B and T lymphocytes, as part of which memory cells are produced. The induced immunity is only effective if there is only one strain of the micro-organism and/or a low mutation rate.

Formation of tissue fluid and lymph

Tissue fluid is the liquid that leaves capillaries and circulates around the cells of an organ. Some returns to the blood, the rest enters **lymph vessels**.

Osmotic force, due to differences in water potential (ψ), and **hydrostatic pressure**, due to contraction of the ventricles, interact to cause the formation of tissue fluid and its return to the blood. All along a capillary, ψ plasma is greater than ψ tissue fluid, so osmotic force tends to draw water in.

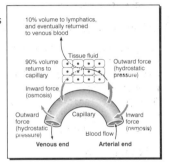

At the arterial end, hydrostatic pressure is greater than the osmotic force; at the venous end it is a smaller force. The balance between the two forces determines the movement of fluid. At the arterial end, the higher hydrostatic pressure forces any molecules that are small enough to pass through the capillary walls; protein molecules are too large. At the venous end, the osmotic force draws water back in to the plasma. Other substances, e.g. carbon dioxide, diffuse in.

Any tissue fluid not returned to the plasma drains into vessels called **lymphatics**. These return it to the blood at the junction of the subclavian vein and the jugular vein. As it returns, it passes through **lymph nodes**, where lymphocytes are added to it. Lymph plays a major part in the transport of lipids. These are absorbed into **lacteals** (lymph vessels in the small intestine) as **chylomicrons**.

Check yourself

1 Explain why a bacterium doesn't need a transport system but a whale does. (4)

2 Describe four differences between arteries and veins. Relate these to the function of the vessels. (8)

3 (a) Copy and complete the table. (6)

	Atrio-ventricular valve (open/closed)	Aortic valve (open/closed)	Blood pressure in ventricle (high/low)
Peak of atrial systole			
Peak ventricular systole			

(b) Describe how the cardiac output is increased during exercise. (4)

(c) Explain how, during exercise, the brain can receive a smaller proportion of the cardiac output yet still receive the same amount of blood. (3)

4 Complete the following paragraph.

The heartbeat is .... (the beat originates within the heart). The .... node produces impulses which travel along .... causing the walls of the atria to .... The only route to the .... is through the .... node. This conducts only .... and so the impulse is held up, before it passes along the .... and causes .... of the ventricles. The delay at the .... node allows time for the .... to empty before the .... contract. The heart rate can be increased by .... impulses from the .... in the medulla and the hormone .... It can be slowed by .... impulses. (16)

5 (a) Explain how the humoral immune response occurs. (6)

(b) Compare active and passive immunity. (3)

6 Describe how differences in hydrostatic pressure and Ψ of the plasma account for the formation of tissue fluid and the return of some of this to the blood. (7)

The answers are on page 110.

Water and dissolved ions are transported upwards from the root in the **xylem**. **Phloem** transports soluble organic compounds, e.g. sugars and amino acids, around the plant. The two tissues are nearly always bound together.

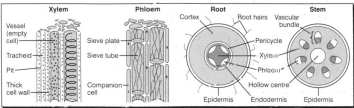

Transpiration

Transpiration is the movement of water through plants. Water enters the roots by **osmosis**. Root hairs increase the surface area available for absorption. The water then moves across the root cortex to the endodermis by either:

- moving through the cytoplasm of adjacent cells (the **symplast pathway**), or
- moving through the cellulose cell walls (the **apoplast pathway**).

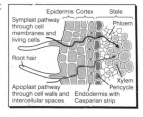

Cells of the endodermis have a waxy **Casparian strip** that blocks the apoplast pathway. From here, water moves to the xylem only by the symplast pathway.

Ions are actively pumped into the xylem. This lowers the water potential (ψ) in the xylem and draws water in by osmosis. As more water enters, it creates a **root pressure** that helps to move the water upwards through the root and stem. The continual stream of water and dissolved ions is the **transpiration stream**.

TRANSPORT IN PLANTS (2)

Although transpiration is continuous, think of it occurring in the following stages.

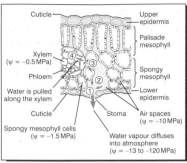

1 Water vapour diffuses from air spaces through open stomata.
2 Water evaporates from the surfaces of spongy mesophyll cells, lowering their ψ.
3 These cells gain water by osmosis from neighbouring xylem vessels.
4 This loss of water creates a negative pressure in the xylem, and water is drawn up the stem.

Factors affecting transpiration

- A humid atmosphere reduces the diffusion gradient between air spaces and the atmosphere, slowing the process down.
- Air movement blows water vapour away and increases the diffusion gradient, so transpiration is quicker on windy days.
- Water vapour molecules have more kinetic energy on hot days, move faster and increase the diffusion gradient, speeding up transpiration.
- A high light intensity usually makes stomata open wider, allowing more water vapour to escape (see page 42).

The cohesion–tension theory

Because of the loss of water vapour from leaves, water in the xylem is under **tension** (negative pressure). Strong forces of **cohesion** hold the water molecules together and so a continuous column of water moves through the plant down a water potential gradient. The gradient is (approximately):

$\psi_{(soil)} = -0.05$ MPa $> \psi_{(root)} = -0.2$ MPa $> \psi_{(stem)} = -0.5$ MPa $> \psi_{(leaf)} = -1.5$ MPa $> \psi_{(air)} = -100$ MPa

(You will not need to remember these values.)

Adaptations of xerophytes

Xerophytes are plants that live in hot, dry conditions. To survive they show one or more of the following adaptations:

- leaves with a much reduced surface area;
- extensive root systems, to absorb as much water as possible;
- swollen stems, which store water;
- surface hairs, which trap moist air near to the leaf surface;
- curled leaves, which maintain a high humidity in the centre;
- sunken stomata;
- restricted stomatal opening times;
- thick waxy cuticle;
- a specialised photosynthesis that allows them to obtain CO_2 at night and fix it into carbohydrate in the day.

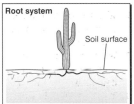

Root system

Soil surface

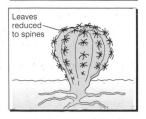

Leaves reduced to spines

Translocation

Translocation is the movement of organic substances through a plant. Phloem transports organic molecules, e.g. sugars, from their site of synthesis (their **source**) to a site of use or storage (a **sink**). The hexose sugars formed in photosynthesis are converted to sucrose for transport.

The pressure flow hypothesis

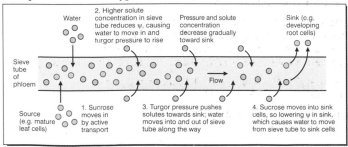

Water

2. Higher solute concentration in sieve tube reduces ψ, causing water to move in and turgor pressure to rise

Pressure and solute concentration decrease gradually toward sink

Sink (e.g. developing root cells)

Sieve tube of phloem

Flow

Source (e.g. mature leaf cells)

1. Sucrose moves in by active transport

3. Turgor pressure pushes solutes towards sink; water moves into and out of sieve tube along the way

4. Sucrose moves into sink cells, so lowering ψ in sink, which causes water to move from sieve tube to sink cells

1 Sucrose is **actively transported** into phloem **sieve tubes** by carrier proteins in the cell membranes. Here it lowers the ψ of the sieve tubes.

2 Water follows the sucrose down a ψ gradient.

3 The extra volume in the phloem creates a high hydrostatic pressure, which forces liquid along the sieve tubes to areas with a lower hydrostatic pressure (the sink).

4 At the sink, e.g. root cells, sucrose is actively 'unloaded', lowering the ψ of the root cells. Water follows down a ψ gradient. The sucrose is rapidly respired or converted to starch for storage, raising the ψ again in the sink cells. The water passes into the xylem and is recirculated to the leaves, maintaining a low hydrostatic pressure.

Cytoplasmic streaming

This theory of translocation suggests that cytoplasm streams round the sieve tubes and passes through the sieve plates from one cell to the next. It explains bi-directional movement in the phloem, but has disadvantages:

● streaming is too slow to move the volumes involved;
● streaming has not been observed in mature phloem.

Cytoplasmic strands

Strands of protein, which are continuous from one sieve tube to the next, passing through the sieve plate, were thought to be tubules that transported organic molecules. They are now thought to help repair damaged phloem cells.

Investigating translocation

In **ringing experiments**, a ring of bark (containing phloem) is removed from a stem. Organic solutes accumulate above the ringed area: the absence of phloem means that they cannot be transported past the ring.

Radioactive tracing, using radioactive carbon or nitrogen, tracks how far, how fast and in which direction organic molecules are moving.

Aphid stylets (mouthparts) pierce stems and suck fluid from the phloem. By cutting off these stylets they can be used as micropipettes obtaining samples of phloem sap for analysis.

1 (a) Describe two similarities between xylem and phloem cells. (2)

    (b) Describe three differences between xylem and phloem. (3)

2 The diagram shows the uptake of water by roots.

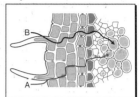

    (a) Which is the symplast pathway and which the apoplast? (2)

    (b) Where is the Casparian strip and what is its function? (3)

    (c) Explain why water enters the root and moves towards the xylem cells in the centre of the root. (3)

3 Complete the following paragraph.

Water is lost from the leaves through open .... The water ...., diffuses out from .... in the spongy mesophyll. This water vapour is replaced by .... from the surfaces of the cells. This .... the water potential of the cells which gain water from nearby xylem vessels by .... This creates a negative pressure or .... in the xylem. Water molecules are pulled .... as a continuous column because of .... of molecules. This is the .... theory of .... (11)

4 List four environmental factors that affect transpiration rate. Explain the effect of each. (8)

5 Explain five adaptations of xerophytes that reduce loss of water. (10)

6 The diagram shows the result of a ringing experiment.

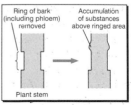

Ring of bark (including phloem) removed   Accumulation of substances above ringed area

Plant stem

    (a) Explain the results in terms of transport in the phloem. (5)

    (b) Describe the pressure flow hypothesis of transport of organic molecules (6)

    (c) How can C^{14} be used to study translocation? (4)

The answers are on page 111.

[not AQA A]

The need for digestion

In **digestion**, enzymes **hydrolyse** large insoluble molecules into smaller soluble ones. These smaller molecules can pass through plasma membranes and into the cells in the gut wall and then to the blood. They can then be:

● **assimilated** – incorporated into structures in the organism;
● **respired** – used as a source of energy to drive metabolic processes.

The human digestive system

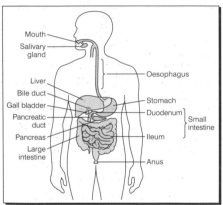

Mouth
Salivary gland
Liver
Bile duct
Gall bladder
Pancreatic duct
Pancreas
Large intestine
Oesophagus
Stomach
Duodenum
Ileum
Small intestine
Anus

The main enzymes involved in digestion in humans are:

● **amylase**, which hydrolyses the amylose and amylopectin in starch into maltose;
● **maltase**, which hydrolyses maltose into α-glucose;
● **lipase**, which hydrolyses lipids, e.g. triglycerides and phospholipids, into glycerol and fatty acids;

● **endopeptidases**, which hydrolyse peptide bonds in the middle of protein molecules;
● **exopeptidases**, which hydrolyse peptide bonds at the ends of protein molecules (carboxypeptidases at the carboxyl end of the molecule, aminopeptidases at the amino end);
● **dipeptidase**, which hydrolyses dipeptides into amino acids.

Region of gut	Secretion	Enzyme(s) & other contents	Digestive action
Buccal cavity (mouth)	Saliva from salivary glands	Amylase Mucus	Starch → maltose None – gives lubrication
Stomach	Gastric juice from gastric glands in stomach wall	Pepsin (an endopeptidase) Hydrochloric acid	Protein → short chain peptides None – provides optimum pH for pepsin and kills bacteria
Lumen of small intestine (duodenum and ileum)	Bile	Bile salts Sodium hydrogen -carbonate	None – emulsify fats, giving larger area for enzyme action Raises pH: inactivates pepsin, gives optimum pH for other enzymes
	Pancreatic juice	Amylase Lipase Trypsin (an endopeptidase) Exopeptidases	Starch → maltose Lipids → fatty acids and glycerol Proteins → short chain peptides Short chain peptides → dipeptides
Cells in wall of ileum		Maltase Dipeptidase	Maltose → α-glucose Dipeptides → amino acids

Pepsin and trypsin are both secreted in an inactive form (pepsinogen and trypsinogen, respectively). This prevents digestion of the cells of the stomach and pancreas (like all cells, their membranes contain protein). Pepsin is activated by hydrochloric acid from the stomach wall and trypsin by enterokinase from the intestinal wall.

Both are endopeptidases and act before the exopeptidases. This creates more 'ends' for the exopeptidases to act on later.

Herbivorous mammals also digest cellulose into β-glucose. To do this, part of their gut is enlarged and contains a large population of micro-organisms. These secrete the enzyme **cellulase**, which digests cellulose. Cattle and other similar animals have a **rumen** (an enlarged part of the stomach) containing the micro-organisms (see page 62). Others, like the rabbit, have an enlarged **caecum** (an elongated part of the intestine).

Absorption of the products of digestion

Most absorption takes place in the **ileum**. The diagram shows how diffusion, facilitated diffusion, active transport and osmosis are involved in the absorption of fatty acids and glycerol, sodium ions, water molecules, glucose and amino acids from the lumen of the ileum.

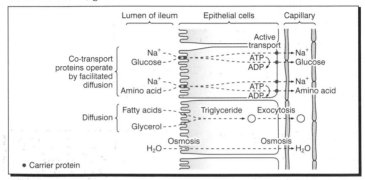

Efficient absorption requires a large area of contact between blood and gut, short transport distances and a steep concentration gradient (unless active transport is involved).

Requirement	Feature of ileum
Large area of contact	Ileum is long, increasing surface area Inner wall has millions of villi Individual cells of epithelium have microvilli Villi contain an extensive capillary network and lymph vessels (to absorb fatty acids and glycerol)
Short transport distance	Lining epithelium is thin Capillaries are close to surface
Steep concentration gradients	Active transport of molecules from epithelial cells to blood maintains low concentration in cells, e.g. glucose and amino acids Movement of blood carries molecules away

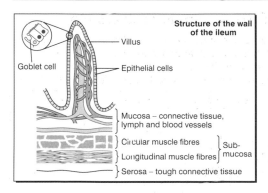

Structure of the wall
of the ileum

Villus

Goblet cell

Epithelial cells

Mucosa – connective tissue, lymph and blood vessels

Circular muscle fibres

Longitudinal muscle fibres } Sub-mucosa

Serosa – tough connective tissue

Controlling digestive secretions

Digestive secretions are controlled by:

- simple reflex actions, which allow a rapid response to the presence of food;
- conditioned reflex actions, which allow a rapid response to the sight, smell or thought of food;
- hormone-controlled reactions, which allow a slower but longer lasting response to the presence of food.

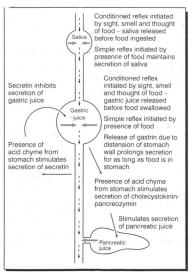

Conditioned reflex initiated by sight, smell and thought of food – saliva released before food ingested

Simple reflex initiated by presence of food maintains secretion of saliva

Secretin inhibits secretion of gastric juice

Conditioned reflex initiated by sight, smell and thought of food – gastric juice released before food swallowed

Simple reflex initiated by presence of food

Release of gastrin due to distension of stomach wall prolongs secretion for as long as food is in stomach

Presence of acid chyme from stomach stimulates secretion of secretin

Presence of acid chyme from stomach stimulates secretion of cholecystokinin-pancreozymin

Stimulates secretion of pancreatic juice

Saliva

Gastric juice

Pancreatic juice

1 **(a)** What is digestion? (3)
  (b) Why is digestion of food necessary in humans? (2)

2 Copy and complete the table. (10)

Region of gut	Secretion	Enzyme(s) & other contents	Digestive action
Buccal cavity	Saliva	Amylase	....
Stomach	....	....	Protein → short chain peptides
		Hydrochloric acid	....
Lumen of small intestine	Bile	Bile salts	....
		....	Provides alkaline pH
	Pancreatic juice	Lipase	....
		....	Starch → maltose
		Trypsin	Proteins → short chain peptides
		Exopeptidases	Short chain peptides → dipeptides
Cells in wall of ileum		Maltase	....
		....	Dipeptides → amino acids

3 **(a)** Why are pepsin and trypsin secreted in an inactive form. (3)
  (b) Why are endopeptidases secreted earlier in the digestive system than exopeptidases? (3)

4 **(a)** Give three features of the ileum that adapt it to efficient absorption. (3)
  (b) Explain how facilitated diffusion and active transport contribute to the absorption of glucose from the ileum. (4)

5 **(a)** Explain the benefit of the secretion of saliva being controlled by both a simple reflex action and a conditioned reflex action. (2)
  (b) Explain the role of cholecystokinin in the control of the secretion of pancreatic juice. (3)

PATTERNS OF NUTRITION (1)

[Edexcel and WJEC only]

Autotrophic nutrition involves an organism **synthesising** (making) organic molecules from inorganic ones. Plants that photosynthesise use light energy to drive the process; they are **photo-autotrophs**. Nitrifying bacteria oxidise ammonium or nitrite to supply the chemical energy they need to synthesise organic molecules; they are **chemo-autotrophs**.

Heterotrophic nutrition involves an organism taking in organic molecules from the environment. There are several different types of heterotrophic nutrition.

Holozoic nutrition involves ingesting parts, or all, of the bodies of other organisms and digesting them, prior to absorption. Animals show this type of nutrition and a specialised gut is needed. **Herbivores** feed on the bodies of plants and **carnivores** feed on the bodies of animals.

Saprobiontic nutrition involves secreting enzymes to digest components of already dead organisms. The products of digestion are then absorbed. The decomposing bacteria and fungi feed in this way.

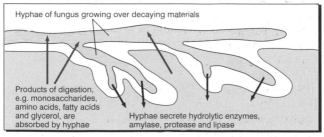

Hyphae of fungus growing over decaying materials

Products of digestion, e.g. monosaccharides, amino acids, fatty acids and glycerol, are absorbed by hyphae

Hyphae secrete hydrolytic enzymes, amylase, protease and lipase

Parasitic nutrition involves a permanent association between two organisms. One (the **parasite**) obtains food from the other (the **host**). The parasite benefits from the association while the host is harmed. **Ectoparasites**, such as fleas, live *on* their hosts whereas **endoparasites**, such as tapeworms, live *in* their hosts.

Mutualistic nutrition involves a permanent association between two organisms in which both benefit from the association, e.g micro-organisims in the rumen of cows.

PATTERNS OF NUTRITION (2)

Herbivorous mammals

Herbivores feed on plants and physically break down tough plant cell walls and digest the cellulose in them. Their teeth and jaws are adapted for this.

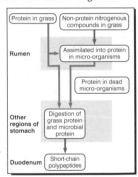

'Sideways' action of jaw grinds vegetation as ridges on upper and lower teeth pass over each other

Diastema

Horny pad

Incisors cut against horny pad

Chisel-like incisors

Ridged premolars   Ridged molars

Most herbivores have a region of the gut containing billions of micro-organisms that digest cellulose. In sheep and cattle, this is a region of the stomach called the **rumen**. Such animals are called **ruminants**. The micro-organisms use compounds containing nitrogen to make protein that the animal cannot use. This extra source of protein is passed on to the animal when micro-organisms pass from the rumen into the next region of the gut and are digested.

This is an example of mutualism, as the micro-organisms receive a supply of plant material to digest as well as being protected in the rumen. The ruminant receives an increased protein supply when the micro-organisms are digested.

Rumen	Protein in grass	Non-protein nitrogenous compounds in grass
		Assimilated into protein in micro-organisms
		Protein in dead micro-organisms
Other regions of stomach		Digestion of grass protein and microbial protein
Duodenum		Short-chain polypeptides

Carnivorous mammals

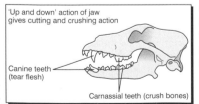

'Up and down' action of jaw gives cutting and crushing action

Canine teeth (tear flesh)

Carnassial teeth (crush bones)

Carnivores have no specially adapted regions of the gut, but their teeth are adapted to cut and slice flesh as well as to crush bones.

1 (a) What is the difference between autotrophic nutrition and heterotrophic nutrition? (2)

(b) Give **two** different examples of autotrophic feeders and **two** different examples of heterotrophic feeders. (4)

2 Complete the following paragraph.

Parasitism and saprobiosis are both example of ....
Parasites can either live in their .... and are called ....
or they can live on their .... and are called ....
All parasites obtain .... and benefit from the association with
their .... Saprobionts are organisms of .... They
decompose the .... bodies of other organisms by secreting
.... They then absorb the products of ....
(11)

3 (a) What is meant by holozoic nutrition? (2)

(b) Give three differences between the teeth of a herbivore and those of a carnivore. (3)

(c) How does the jaw action of a herbivore and a carnivore differ? (2)

The answers are on page 113.

ECOLOGY (1)

[not AQA A or AQA B]

Ecological terms

Ecosystem	A self-supporting system of organisms (producers, consumers, decomposers) interacting with each other and with their physical environment. A garden pond is a small ecosystem.
Community	The sum of all the organisms in *all species* in an ecosystem at a given moment.
Population	All the individuals of *one species* in an ecosystem at any moment.
Environment	The sum of the conditions/factors that affect the success of an organism. They are of two main types. **Biotic factors** Factors due to other organisms, e.g. predation by another species or infection by a micro-organism). **Abiotic factors** Factors due to non-living aspects of the environment, e.g. physical structures, availability of water and mineral ions, light intensity, temperature.
Habitat	The area in an ecosystem where a particular species is found. For example in a pond the habitat for most fish is the open water, for pondskaters it is the surface of the water, for snails it is mainly the soil surface at the sides of the pond.
Niche	The role an organism fulfils within a given habitat. If two animal species feed from the same plant, but at different times or from different parts of the plant, they have different niches and will not compete with each other.

Feeding relationships in ecosystems

A **food chain** is the simplest way of showing feeding relationships. For example:

Grass (Producer)	→	Rabbit (Primary consumer) (Herbivore)	→	Fox (Secondary consumer) (Carnivore)

The different stages in a food chain are called **trophic levels**. Food chains rarely have more than five trophic levels (and usually have fewer) because of energy losses between the trophic levels (see later).

Food chains in an ecosystem usually interact to form **food webs**.

ECOLOGY (2)

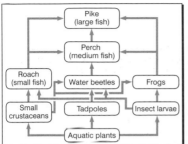

Changes in the population of one organism can influence those of another even if they are not directly linked in a food chain. For example, if there are fewer water beetles, this can cause a decrease in the population of roach because perch will eat more roach to make up for the lack of water beetles.

Food chains can be represented diagrammatically in:

● **pyramids of numbers**, which represent the total numbers of organisms at each trophic level at a moment in time, irrespective of their mass;
● **pyramids of biomass**, which represent the total biomass of organisms at each trophic level at a moment in time, irrespective of their numbers.

For the food chain **plankton → crustaceans → small fish → pike** the pyramids are:

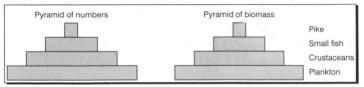

The organisms increase in size along the food chain and, quite logically, the numbers decrease. The total biomass also decreases.

For the food chain **oak tree → aphids → ladybirds → blackbird** the pyramids are:

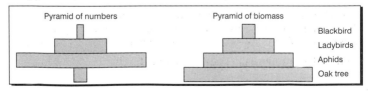

Because of the large size of an oak tree, it can support thousands of the much smaller aphids. From this point in the chain, the size of organisms increases and the numbers decrease. Again, the total biomass decreases along the food chain. Biomass decreases along a food chain because:
- some parts of an organism are not eaten, e.g. roots, bones;
- some parts that are eaten are indigestible and pass out in faeces;
- some materials are respired and the CO_2 formed is lost to the environment;
- metabolism forms other excretory products that are lost to the environment.

Energy transfer in ecosystems

Energy transfer diagrams represent the total flow of energy through an ecosystem over a period of time. Key processes in energy transfer include:
- **photosynthesis**, which fixes light energy into bonds between organic molecules;
- **respiration**, which releases energy from organic molecules that is used to make ATP which can the 'drive' other processes;
- **synthesis** of molecules and cells, which uses energy that may be passed on to the next trophic level in the food chain when the organism is eaten;
- **biological** processes, which use energy, e.g. muscle contraction and active transport; the energy is lost to the environment and cannot be passed to the next trophic level.

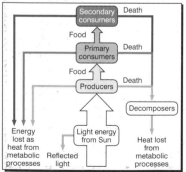

Over a period of time, all the light energy that is used by producers in photosynthesis will pass through the ecosystem in the ways shown. In any one year, some energy will remain in the new growth of the organisms in the ecosystem.

Cycling carbon and nitrogen through ecosystems

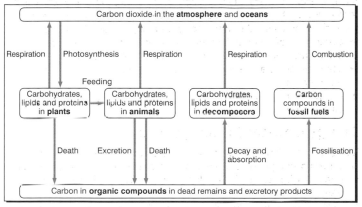

In the **carbon cycle**, photosynthesis and respiration influence the levels of CO_2 in the atmosphere. In summer, with more leaves on trees and longer days, the rate of photosynthesis is greater than the overall rate of respiration. More CO_2 is used than is released and levels in the atmosphere fall. In winter, the situation is reversed and levels rise.

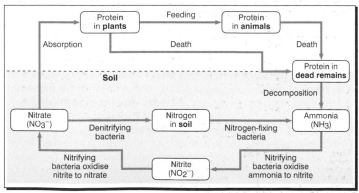

In the **nitrogen cycle**, don't link the cycling of nitrogen too closely with energy transfer. When plants absorb nitrates, any energy in the nitrates is not available to the plants to drive metabolic processes.

Human impact on the environment

Global warming

Evidence suggests that the Earth's temperature has been rising for some time. This could be due to deforestation and increased combustion of fossil fuels enhancing the greenhouse effect. **Global warming** could lead to:

● increased sea levels due to melting of polar ice caps;
● long-term climate change – more rainfall overall, but less in some areas;
● change in ecosystems as organisms better adapted to the changed conditions enter and multiply faster than existing species;
● extinctions as new species out-compete existing ones that are unable to colonise new areas quickly enough.

Deforestation

Tropical rainforests comprise not just the huge trees, but also millions of other animals, plants and decomposers. A rainforest is a complex ecosystem and when trees are felled, there are many possible effects on the CO_2 levels, including:

● loss of trees, which absorb CO_2 from the atmosphere;
● loss of animals, which add CO_2 to the atmosphere;
● burning the trees, which adds CO_2 to the atmosphere.

Deforestation has other effects, including:

● increasing soil erosion, as the shallow soil in rainforest is protected from winds by the tree canopy and from the effects of flooding by the tree roots;
● reducing species diversity, as many ecological niches are lost;
● reducing the nitrate content of soil, as most nitrate absorbed from the soil remains fixed in proteins and other molecules in the huge tree trunks that are slow to decay and may even be removed from the area.

Greenhouse effect

'**Greenhouse gases**' in the atmosphere reduce the amount of heat escaping from the Earth to space. This has always happened and is the reason why the Earth is warm enough to sustain life. Increasing levels of some gases (CO_2, methane and CFCs) have increased this effect and the Earth's temperature is rising.

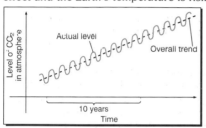

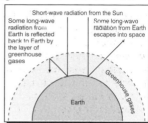

Reduction of ozone layer

Ozone is a form of oxygen in which three atoms combine to form a molecule (O_3). The **ozone layer**, reduces the amount of harmful ultraviolet radiation (UV) reaching the Earth from the Sun. Gases like CFCs react with ozone converting it to oxygen (O_2). More UV can then reach the Earth and this may result in an increase in skin cancers. One molecule of a CFC can decompose many thousands of molecules of ozone.

Water pollution

Two forms of **water pollution** are especially significant.

Nitrates are extremely soluble and, when applied to land in the form of inorganic fertilisers, dissolve in rain water and may be leached from soils into nearby waterways, causing **eutrophication**.

Organic pollution (by manure, sewage, etc.) allows decomposers to multiply rapidly, reducing the levels of dissolved oxygen as they use it in respiration. Animals like blood worms and rat-tailed maggots tolerate low oxygen levels and are **indicator species** of organic pollution.

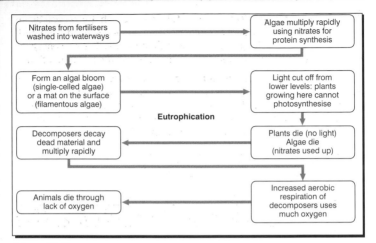

| Nitrates from fertilisers washed into waterways | → | Algae multiply rapidly using nitrates for protein synthesis |

Eutrophication

Air pollution

Combustion of fossil fuels raises levels of sulphur dioxide (SO_2) and nitrogen oxides (NO_x). These gases can be carried hundreds of miles by winds and then dissolve in rain water to form acid rain, which can:

● kill trees directly, especially conifers;
● acidify waterways and kill or limit reproduction of fish and other animals;
● release toxic mineral ions, e.g. aluminium Al^{3+}, from soils into water (Al^{3+} is more soluble at low pH).

The levels of SO_2 can be monitored by observing the types of lichens growing in an area. They act as indicator species; shrubby (bushy) lichens will only grow where SO_2 levels are low, whereas crusty (flatter) lichens tolerate higher levels of pollution.

Overuse of pesticides

A **pest** is an organism that reduces the yield of a crop plant or stock animal. **Pesticides** kill, or limit the reproduction of, pests and include insecticides (such as DDT) to kill insect pests, herbicides (weedkillers) to kill weeds, and fungicides to kill fungi.

Pesticide resistance arises when mutations occur that give resistance to a pesticide. The resistant organisms have a selective advantage and survive to reproduce in greater numbers than others. The percentage of resistant organisms in the population increases with each generation.

Some pesticides are not **biodegradable** and persist in the environment. They enter organisms that were not their target, are stored in fatty tissue and are passed along food chains. This is **bio-accumulation**. Levels rise along the food chain as, typically, many organisms of one level are eaten by just one of the next.

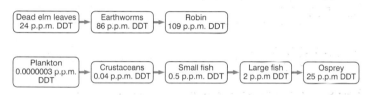

Succession and species diversity

Ecosystems change with time, becoming more complex until a **climax community** is established. This is the most stable community of dominant species that can exist under these conditions, and usually includes trees. The stages in a succession are called **seres**.

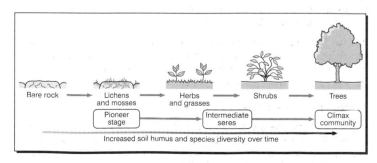

The changes are not restricted to the plants. As more and varied plant species enter, they create more niches for animals. Both change the physical environment. Many factors can determine the climax, such as rainfall, temperature and grazing. As the community becomes more complex, the **species diversity** increases. Species diversity measures the range of species and their success in an ecosystem. It is more than just the number of species in an area. **Simpson's diversity index (d)** is calculated from the formula: $d = \dfrac{N(N-1)}{\Sigma n(n-1)}$

N = total number of organisms in the area

n = number of organisms of a particular species in the area

Σ = sum of

A high diversity index suggests a stable ecosystem with complex food webs. A low index suggests a less stable ecosystem dominated by just a few species.

Estimating population size

To estimate the size of an animal population in an area:
- collect a random sample of the animals and record the number (N_1);
- mark the animals unobtrusively with a non-toxic paint and release them;
- allow time for the organisms to distribute evenly in the population, collect and count a second sample (N_2);
- record the number of marked animals in this sample (n);
- estimate population size as $\dfrac{N_1 \times N_2}{n}$.

This technique assumes that there is no change in the total numbers (no deaths, no reproduction, no migration) of the population during the time between collecting the two samples.

To estimate the size of a plant population in an area:
- estimate the area of the region being investigated (X);
- throw quadrats of a known area (Y) at random and count the numbers of plants in each quadrat; obtain a mean number per quadrat (Z);
- estimate the population size as $\dfrac{X \times Z}{Y}$.

Distribution of organisms

One method of investigating the distribution of organisms in an area is the **belt transect**:
- lay a measure across the area and determine regular sampling points, e.g. every 5 metres;
- at each point lay five quadrats next to each other;
- estimate the abundance of each species in each quadrat and obtain a mean value for that sampling point.

Adaptation to environment

Plants adapted to very dry conditions are called **xerophytes** (see page 53 for adaptations).

Mammals living in hot deserts, e.g. kangaroo rats, often:
- produce small mounts of concentrated urine;
- lose heat effectively, e.g. through large ears;
- show temperature tolerance (do not sweat as soon as body temperature rises).

They may also be nocturnal and so avoid the most intense heat.

Mammals living in cold environments, e.g. polar bears, typically:
- have thick fur and subcutaneous fat to provide insulation;
- are large, as a reduced surface area to volume reduces heat loss.

1 **(a)** What is an ecosystem? (3)

    (b) What is the difference between:

        (i) a population and a community? (2)

        (ii) a habitat and niche? (2)

        (iii) biotic and abiotic factors? (1)

2 For the food chain *plankton → crustaceans → fish → killer whale*

    (a) Name the producer and secondary consumer. (2)

    (b) Draw a pyramid of biomass of the food chain (2)

3 Complete the following paragraph.

Not all the light energy striking a plant is used in .... Some of it is .... back into space, some is the wrong .... to be used. Not all of the energy fixed in .... by .... is passed on to the next .... level. When the plant dies, some passes to the .... At each level, only about .....% of the energy entering that level is passed to the next. This limits the number of levels in a .... as there is just not enough .... to support extra levels. (10)

4 Copy and complete the diagram of the carbon cycle. (4)

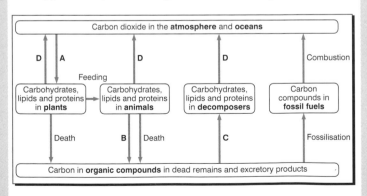

The answers are on page 113.

5 **(a)** Explain why levels of CO_2 in the atmosphere:
   (i) fluctuate in any one year; (5)
   (ii) have increased overall during the last 100 years. (4)
   (b) State and explain three consequences of global warming. (6)

6 **(a)** **(i)** Why can an increase in nitrates in a pond cause an algal bloom? (3)
   (ii) How can an algal bloom lead to an increase in the numbers of decomposers in the pond? (2)
   (iii) How does this decrease the oxygen levels in the pond? (3)
   (b) Why does organic pollution of water decrease the oxygen levels? (3)

7 How would you estimate the population of:
   (a) plants in an area? (4)
   (b) animals in an area? (4)

8 **(a)** State and explain three adaptations of xerophytes to their environment. (6)
   (b) State and explain two adaptations shown by mammals to:
   (i) a hot, dry environment; (4)
   (ii) a cold environment. (2)

The answers are on page 113.

(Edexcel only)

The structure of a flower

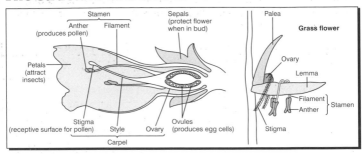

You must be able to name all of the parts and describe their functions.

Gamete formation

Male **gametes** are contained in the **pollen grains**, formed in the **anthers**; each pollen grain contains two gametes. They are not fully formed until after pollination. Female gametes are formed in the **ovules** of the **ovaries**.

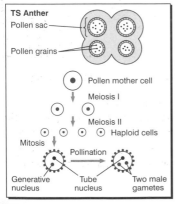

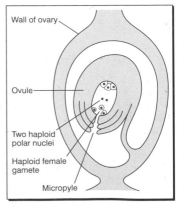

Pollination

Pollination is the transfer of pollen from the anther to the **stigma**, usually in a different flower. **Wind** and **insect** pollination are the two most common methods.

Feature	Insect-pollinated plant	Wind-pollinated plant
Petals	Large (in proportion to flower), colourful, scented to attract insects	Small, anthers and stigmas exposed to wind
Nectaries	Present, nectar ensures more visits by insects to others of same species	Absent
Stigmas	Usually compact, placed to touch the insect's body when in flower	Often 'feathery', giving large surface area to 'catch' air-borne pollen grains
Pollen grains	Often sticky or with small hooks to attach to insect's body	Often small, light with extensions to increase surface area for lift

Fertilisation and seed development

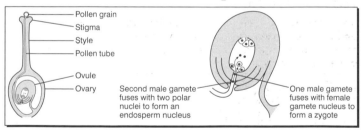

Self-fertilisation is prevented by a range of mechanisms including:
● separate male and female flowers;
● stamens and ovaries maturing at different times;
● self-incompatibility, e.g. pollen tubes will not grow on stigmas of the same species or, if fertilisation does occur, zygote fails to develop.

Seeds develop from ovules that contain fertilised egg cells; the ovary becomes a **fruit**. Biologically a pea pod is a fruit containing several pea seeds.

1 (a) Give the names of parts A – F on the diagram of a typical flower. (6)

(b) Give the letter of the part that will:

   (i) attract insects; (1)

   (ii) become a fruit. (1)

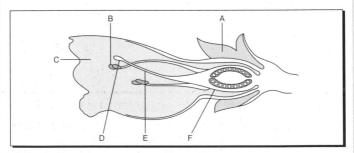

2 (a) What is meant by pollination? (2)

(b) Give and explain three differences between insect-pollinated flowers and wind-pollinated flowers. (6)

3 Complete the following paragraph.

After pollination, the .... grows a .... which penetrates the style and grows down towards an .... The nuclei from the .... pass down the .... and when they reach the ...., the two male .... pass into the .... Here, one fuses with the two .... nuclei and the other fuses with the female gamete to produce a .... (10)

4 State and explain three ways in which self-fertilisation is prevented. (6)

5 (a) Explain the difference between a seed and a fruit. (2)

(b) State three ways in which seeds/fruits may be dispersed. (3)

The answers are on page 114.

[Edexcel Human Biology only except where indicated otherwise]

Human reproductive systems

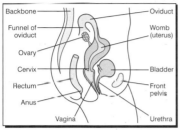

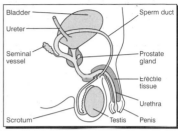

Human gametogenesis (gamete formation)

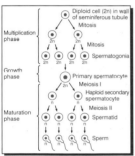

The male sex cells are the **spermatozoa** (sperm). Sperm are produced from **epithelial cells** lining the **seminiferous tubules** of the **testes** by a combination of mitosis and meiosis. Meiosis produces genetic variation in the mature sperm.

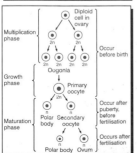

A similar set of processes forms the female sex cells. However, gametogenesis in females is put 'on hold':
- after the multiplication phase, which occurs before birth, until puberty;
- after the first meiotic division (during each menstrual cycle one oocyte develops to this stage).

The second meiotic division only takes place if a sperm nucleus enters the oocyte.

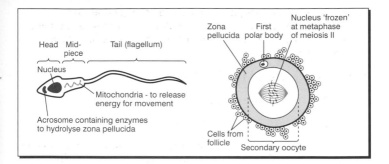

Fertilisation

One sperm makes contact with the **zona pellucida**. The **acrosome** of this sperm releases enzymes that hydrolyse the glycoprotein, this is the **acrosome reaction** which creates a path through the zona pellucida. The sperm passes through and its membrane fuses with that of the oocyte. The sperm nucleus (the male gamete) enters the oocyte, stimulates the second meiotic division by the oocyte, which then becomes an ovum. The sperm nucleus and ovum nucleus fuse and a **zygote** is formed.

Implantation and early development

The zygote will divide repeatedly by mitosis and form a **blastocyst**, a hollow ball of cells with an increased cell mass (the potential embryo) at one end. The blastocyst will **implant** in the lining of the uterus where development will continue.

Embryonic tissue will form most of the **placenta**, through which exchange of nutrients, respiratory gases and excretory products will occur. Exchange across the placenta is efficient because:

● the folded surface membrane of the placenta creates a **large surface area**;
● the surface membrane is very thin and is surrounded by pools of maternal blood, creating a **short diffusion pathway**;
● fetal and maternal circulations maintain **concentration gradients** across the surface membrane so that diffusion/facilitated diffusion is continuous.

In the later stages of pregnancy, antibodies may pass across theplacenta from mother to fetus. The placenta can also allow some harmful chemicals to pass from mother to fetus, including alcohol, nicotine and heroin.

Hormonal control of reproduction

Control of the menstrual cycle

[AQA A Biology and Edexcel Human Biology]

Oocytes are contained in groups of cells called **follicles**. Two pituitary hormones, **FSH** and **LH**, stimulate follicles to develop and release **oestrogen** and **progesterone**. Changing levels of these hormones produce the events of the menstrual cycle.

Changes in	Days 1–4	Days 5–13	Day 14	Days 15–22	Days 23–28
Uterus lining	Breaks down, menstruation	New lining develops and thickens	Continues to thicken and vascularises	Continues to vascularise	Starts to break down, but no menstruation
Follicles	Enlarges, secretes oestrogen	Continues to develop and secrete oestrogen	Ovulation: egg cell is released from follicle	Remains form corpus luteum and secrete progesterone	Corpus luteum shrinks
FSH levels from pituitary	Low level, increasing	Increases then decreases due to inhibition by oestrogen	Sharp rise	Decreases, inhibition by oestrogen and progesterone	Levels remain low
LH levels from pituitary	Low levels	Levels begin to increase	Levels peak	Decreases, inhibition by oestrogen and progesterone	Levels remain low
Oestrogen levels	Low levels	Levels peak on days 12–13	Levels falling	Levels low and falling	Levels low
Progesterone levels	Low levels	Low levels	Levels low	Levels rising as corpus luteum is active	Falling as corpus luteum is regressing

Hormone	Secreted by	Target	Effect
FSH	Pituitary gland	Cells in follicles	Stimulates development of the follicles and secretion of oestrogen from follicles
LH	Pituitary gland	Cells in follicles	Stimulates ovulation (corpus luteum forms from remains of follicle) and secretion of progesterone by corpus luteum
Oestrogen	Cells in follicle in ovary	Uterine lining Pituitary gland	Stimulates development of uterine lining Inhibits secretion of FSH
Progesterone	Cells in follicle/ corpus luteum	Uterine lining Pituitary gland	Maintains and vascularises uterine lining Inhibits secretion of FSH and LH

Control of the birth process

Throughout pregnancy, the placenta produces oestrogen and progesterone in increasing amounts. These inhibit the secretion of FSH and LH and so prevent further ovulation. Progesterone also inhibits contractions of the muscle of the uterus. Just before birth, the levels of oestrogen and progesterone drop and the pituitary gland secretes another hormone, **oxytocin**. As a result:

● smooth muscle in the uterus wall begins to contract;
● contractions are made stronger by oxytocin, until the **amnion** breaks releasing **amniotic fluid**, marking the onset of **labour**;
● first stage of labour, the cervix dilates;
● second stage of labour, contractions expel the baby;
● third stage of labour, contractions expel the placenta and some maternal tissue (the afterbirth).

Following birth, the baby will suckle the mother's nipple to obtain milk. Milk is produced in the **mammary glands** of the breasts when they are stimulated by the hormone **prolactin** from the pituitary gland. Suckling encourages the production of **oxytocin**, which causes the expulsion (let down) of the milk into the ducts leading to the nipple.

Reproduction in other mammals [AQA A Biology only]

Other female mammals have similar reproductive cycles to humans but without menstruation (only apes menstruate). These are called **oestrous** cycles. The period in the cycle when ovulation occurs and fertilisation may take place is called **oestrus** (note the difference in spelling). Farmers talk of their animals 'coming on heat', because the body temperature rises. At oestrus, cows become restless and allow themselves to be mounted by others in the herd and also try to mount others.

Control of human fertility [AQA A Biology only]

Female infertility is treated by using FSH or **analogues** (molecules similar to FSH). Injections of FSH cause several follicles to develop that are removed from the ovaries before the oocytes are mature. The oocytes are maintained in a liquid medium and sperm from the partner are added. They are incubated and examined for evidence of fertilisation. This is called **in vitro fertilisation (IVF)**. Two or three embryos at the four- or eight-cell stage of development are implanted in the lining of the uterus.

Controlling fertility in domestic animals

[AQA A Biology only]

IVF in domestic animals is usually used to produce many offspring.
- Injections of FSH induce multiple ovulation in a 'superior' cow.
- The oocytes are inseminated with sperm from a 'superior' bull.
- The embryos are selected for gender and at the four- or eight-cell stage are split into single cells again.
- These cells behave like zygotes and develop into more embryos.
- The embryos are implanted into 'surrogate' mother cattle.

Progesterone is used to synchronise breeding behaviour in sheep.
- Sheep are injected with progesterone for several days: this inhibits secretion of FSH and LH and 'suspends' the oestrous cycle.
- The injections are withdrawn: the inhibition of FSH and LH ceases and all the sheep start a new cycle at the same time.

All the sheep can be inseminated at the same time.

Check yourself

1 Give **two** similarities and **three** differences between gametogenesis (gamete formation) in males and females. (5)

2 Complete the following paragraph.

   Fertilisation occurs when the nucleus of a sperm fuses with that of an .... Before this can happen, the sperm nucleus must enter an .... (it is not yet an .... as the second .... division has not yet taken place). The acrosome in the 'head' of the sperm releases .... enzymes which digest the .... which makes up much of the .... This creates a .... through which the sperm can pass. The .... of the sperm fuses with that of the .... and the sperm nucleus enters. The second .... division occurs in the .... which becomes an .... The sperm nucleus fuses with the .... nucleus to form a .... (15)

3 **(a)** What is a blastocyst? (3)
   (b) Explain three features of the placenta that make it efficient at transferring substances between mother and fetus. (6)

4 The diagram shows the levels of hormones during the menstrual cycle.
   (a) Identify, with reasons, the hormones. (8)
   (b) What event occurs when levels of hormones A and B peak? (1)
   (c) What event occurs when the levels of hormones C and D fall? (1)

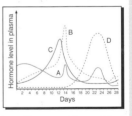

5 Explain how, just before birth, the following effects are brought about.
   (a) Initial contractions of the uterus that break the amnion. (2)
   (b) Breakdown of the placenta. (3)

6 **(a)** What is IVF and why is it used in humans? (3)
   (b) Explain how breeding behaviour can be synchronised is sheep. What is the benefit of this to a farmer? (6)

The answers are on page 115.

[AQA A Biology (Human) and Edexcel Biology (Human)]

Growth and development

Growth is a permanent increase in the amount of living tissue of an organism. In humans, this also means an increase in the number of cells. Growth can be measured by measuring changes in the following.

● **Body mass**. Over a period of time this will give a good indication of the amount of extra tissue but, in the short term, a sudden increase may be due to drinking and a decrease due to dieting/loss of excess fluid.

● **Height**. This is a convenient measure of growth, but takes no account of increase in width/girth.

● **Supine length** (length from head to toe when lying face down). This gives a slightly larger value than height. When standing, gravity compresses the body slightly.

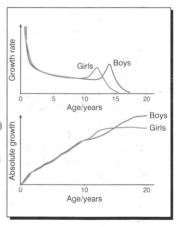

You must be able to distinguish between **growth rate** and **absolute growth**. The growth rate is the amount of extra growth per unit time (usually per year). The absolute growth is usually shown as the total body tissue (total mass or total height) at any one time.

In humans, after the age of 2 growth proceeds at a more or less uniform rate until **puberty**. At this stage, increased secretion of reproductive hormones stimulates an increase in the secretion of growth hormone, which causes the **adolescent growth spurt**. This generally happens earlier in girls than in boys.

Tissues and organs grow at different rates.

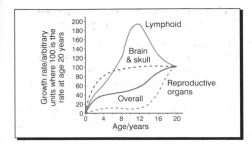

The head and brain reach full size by about age 5–6 and further development is concerned with increased nerve connections allowing ever more complex tasks.

Lymphoid tissue develops rapidly in childhood and adolescence; this is linked to combating disease and developing immunity. Examples of lymphoid tissue include the thymus gland and tonsils.

Reproductive organs only develop fully after puberty, when reproductive maturity is reached.

Ageing

Ageing has many effects. By the age of 70 the following features have decreased performances compared with early adulthood:

● cardiac output (65% of maximum);
● respiratory capacity (55% of maximum);
● basal metabolic rate (80% of maximum);
● nerve impulse conduction velocity (85% of maximum).

Other changes include the following.

Reproductive capacity of both sexes declines. Men produce fewer sperm, although men of over 70 have fathered children. Women cease ovulating at around the age of 45–50. This is the **menopause** and occurs because the follicles in the ovaries no longer respond to FSH from the pituitary. They no longer develop and no longer secrete oestrogen and progesterone. Menstruation ceases as a result. Some of the effects of the menopause can be alleviated in many women by the use of **hormone replacement therapy (HRT)**, which replaces the missing oestrogen.

The **cardiovascular system** (heart and associated blood vessels) becomes less efficient, partly due to the effects of **atherosclerosis**. Fatty deposits are laid down in the lining of arteries as we age. This reduces the diameter of the arteries and makes the passage of blood more difficult. The heart must pump harder as a result and blood pressure increases. Arteries may become blocked by atherosclerosis or by a blood clot (**thrombus**) forming in a narrowed artery. This condition is called **thrombosis** and if it occurs in a coronary artery (in the heart muscle) it is **coronary thrombosis** and may cause a heart attack.

As we age, two conditions of the bone become more common.

Osteoporosis or **brittle bone disease** is a condition in which increasing amounts of bone mass are lost as some of the bone cells are not replaced and the bone becomes progressively less dense. In men, the density usually remains above the fracture threshold, but in post-menopausal women the density can fall below this level.

This is because one of the effects of oestrogen is to encourage cell division in bone tissue to replace cells that are lost. After the menopause, the levels of oestrogen are too low to encourage cell replacement. HRT helps to reduce osteoporosis in post-menopausal women.

Osteoarthritis is evident in more than 80% of people aged 65 or more. It is often caused by repeated stress on a particular joint. For example, some manual workers are susceptible to arthritis of the hip joint because it is continually stressed in their work. The condition develops in the following way:
● repeated stress damages the cartilage at a joint;
● cartilage is lost from the articulating surfaces of the joint;
● there is an increase in the bone-forming cells (**osteoblasts**) in these bones;
● new bone can be formed that produces a deformed shape and makes movement more difficult and painful.

1 (a) Explain the difference between relative growth rate and absolute growth. (2)

(b) Explain the differences in the growth rates of males and females shown in the graphs. (5)

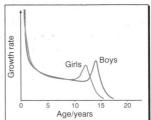

2 The graph shows the growth of different tissues from birth to adulthood. Identify, with reasons, the lines which represent:

(a) growth of lymphoid tissue; (2)

(b) growth of the head. (2)

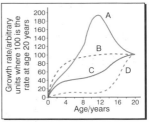

3 (a) Complete the following paragraph.

As we age, the reproductive capacity of both sexes declines. Men produce fewer ...., but can often still father children at the age of .... At around the age of ...., women enter the .... Follicles in the ovary are no longer sensitive to the pituitary hormone .... and so they do not develop into mature .... Levels of the hormones .... and .... decrease sharply and .... ceases as a result. Hormone replacement therapy (..................) can alleviate some of the effects of the .... by replacing the missing .... (12)

(b) List four other effects of ageing. (4)

4 (a) Describe the role of oestrogen in treating osteoporosis. (3)

(b) Explain how repeated stress on a joint can lead to the development of osteoarthritis. (4)

The answers are on page 116.

[AQA A Biology (Human) and OCR Biology only]

Disease is a condition in which the body, or part of the body, does not function normally. A **pathogen** is an organism (usually a micro-organism) that causes disease in another organism. Micro-organisms of all types can cause disease. For example:

- certain **bacteria** cause pulmonary tuberculosis (TB), cholera and food poisoning;
- certain **viruses** cause AIDS, influenza and measles;
- certain **fungi** cause athlete's foot and other skin complaints;
- certain **Protoctista** cause malaria and sleeping sickness.

Other types of organism that cause disease include **tapeworms** and **roundworms** (nematode worms).

Growth of bacterial populations

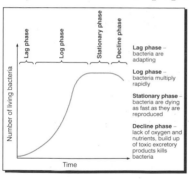

Bacteria are single-celled **prokaryotes** that reproduce mainly by **binary fission**. Each bacterial cell grows, reaches full size then divides into two cells. The time taken for this is the **generation time**. Under favourable conditions this can be as little as 20 minutes. A graph of the growth of a bacterial population shows a typical **sigmoidal** ('S' shaped) curve, with four distinct phases.

The growth of a population is influenced by temperature because all metabolic processes speeded up with an increase in temperature, up to the optimum. The population increases more rapidly with increasing temperature to the same **plateau level**.

Bacteria require nutrients for the synthesis of cytoplasm and organelles as well as for respiration. If nutrients are not freely available, the bacteria will compete for those present and the increase in the population will be less rapid. It may decline if nutrients are in very short supply.

Most bacteria are aerobic, and a lack of oxygen will limit respiration and the release of energy. Growth will be slowed.

The numbers of bacteria growing in liquid medium can be estimated using a **haemocytometer**. This is a special slide with a cavity with a grid marked on the glass. With the cover slip in place, each square holds a known volume of liquid. By counting the numbers of cells in the squares and finding an average, the number of bacteria per cm^3 can be estimated.

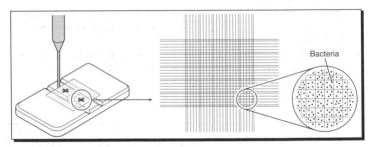

Bacteria

Bacteria and disease

Bacteria cause disease by releasing **toxins** or damaging cells. Bacterial **exotoxins** are secreted by the bacteria as they grow. **Endotoxins** form part of the bacterial cell wall and are released when the bacterium dies.

Bacteria enter the body by one of four main routes:
● breathed in with the air in tiny droplets containing thousands of bacteria;
● through breaks in the skin;
● during sexual intercourse;
● in contaminated food or water.

This process is called **infection**.

Disease does not automatically result just because we have been infected. The bacteria must multiply to a significant level before the amount of toxin released can cause any harm. During this time, the immune system may detect them and destroy them.

Do not confuse infection (the entry of the pathogen) with disease (the appearance of symptoms, the consequences of the pathogen multiplying).

The progress of a bacterial disease can be linked to the bacterial growth curve.
- During the **lag phase**, the bacteria are becoming established: there are too few to cause disease.
- During the **log phase**, the bacteria are multiplying rapidly and releasing increasing amounts of toxins. They are also stimulating an immune response.
- During the **stationary phase**, the bacteria die as fast as they are being produced. We are still ill because of the large amount of toxins being released.
- In the **decline phase**, the immune system is destroying the bacteria faster than they can reproduce. The level of toxins released decreases and we recover.

Viruses and disease

Viruses are **acellular**: they are not true cells and have no organelles. Most viruses consist of a strand of genetic material (DNA or RNA) enclosed in a protein coat.

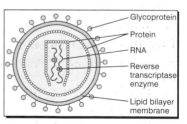

They can only reproduce by invading other cells, where their genetic material directs the host cell's metabolism to produce more viruses.

Diseases caused by pathogenic micro-organisms

AIDS   [AQA A Human Biology and OCR Biology only]

Acquired immune deficiency syndrome (AIDS) is caused by the **human immunodeficiency virus (HIV)**, which infects helper T lymphocytes. The course of infection and reproduction is as follows:

- the virus injects RNA and reverse transcriptase (see page 35) into the cell;
- reverse transcriptase makes a DNA copy of the viral RNA, which becomes part the DNA of the host cell;
- protein synthesis or cell division activates this HIV DNA and it directs the manufacture of HIV proteins;
- some of the proteins are incorporated in the membrane of the helper T lymphocyte, others are used to assemble new viruses;
- the proteins in the membrane stimulate an immune response and the cell is destroyed, releasing the viruses to infect other helper T lymphocytes.

The body produces more helper T lymphocytes, and B lymphocytes produce antibodies against the virus. At this stage, a person is said to be HIV-positive. The cycle of destruction and replacement of helper T lymphocytes can continue for up to 20 years. Eventually, the immune system can no longer keep pace and the number of viruses rises rapidly as the number of helper T lymphocytes decreases. The reduced number of helper T lymphocytes can no longer stimulate enough B lymphocytes to produce antibodies against invading micro-organisms, and so people with AIDS develop bacterial diseases like TB.

Salmonellosis [AQA A Human Biology only]

Bacteria of the genus *Salmonella* cause **salmonellosis** food poisoning. These bacteria are unusual in that they actually enter the epithelial cells lining the small intestine. They cause irritation and the normal uptake of nutrients and water is disrupted, leading to nausea, vomiting and diarrhoea.

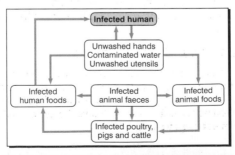

Good personal hygiene, good institutional hygiene and correct cooking can break many of the routes of transmission.

Cholera [OCR Biology only]

Cholera results from an infection by a small comma-shaped bacterium, *Vibrio cholerae*. It is primarily spread by drinking contaminated water, but poor sanitation can lead to food becoming contaminated and routes of transmission very similar to those of salmonellosis. Infection causes severe diarrhoea and vomiting. Rehydration by drinking a solution containing the correct balance of sugars and ions is usually all that is necessary, but a course of antibiotics may be necessary. Cholera is controlled in much the same way as salmonellosis.

Pulmonary tuberculosis (TB)

[AQA A Human Biology and OCR Biology only]

Pulmonary tuberculosis (TB) results from infection by the bacterium *Mycobacterium tuberculosis*. It is normally spread by inhaling air-borne droplets that contain the bacteria. In the lungs, the bacteria multiply and form the small lumps or tubercles that give the disease its name. If the immune system does not destroy the bacteria at this stage, they can spread to other organs and the condition can be fatal.

A different strain of *Mycobacterium* causes TB in cattle and can be spread to humans in milk. In many countries, this route of transmission has been effectively eradicated by:
- pasteurisation of milk, which kills the bacteria;
- regular inspection of cattle and slaughter of the herd where any infected cow is found.

Diseases caused by parasites

Parasites are organisms that gain nutrition from their host. Disease can result from their process of obtaining food, rather than from invading cells or releasing toxin. Parasites are adapted to survive in hostile environments in the host and have life cycles that give a good chance of reinfecting the host.

Malaria [AQA A Human Biology and OCR Biology only]

Malaria is caused by single-celled protoctistans of the genus *Plasmodium*. Different species cause different forms of malaria, but the life cycle of the parasite is similar in all cases.

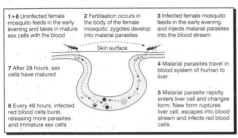

1+8 Uninfected female mosquito feeds in the early evening and takes in mature sex cells with the blood

2 Fertilisation occurs in the body of the female mosquito: zygotes develop into malarial parasites

3 Infected female mosquito feeds in the early evening and injects malarial parasites into the blood stream

Skin surface

7 After 28 hours, sex cells have matured

4 Malarial parasites travel in blood system of human to liver

5 Malarial parasite rapidly enters liver cell and changes form. New form ruptures liver cell, escapes into blood stream and infects red blood cells

6 Every 48 hours, infected red blood cells burst, releasing more parasites and immature sex cells

The female mosquito has a key role in the transmission of the malarial parasite. By feeding in the early evening, she is able to ingest mature sex cells that were released from red blood cells the previous afternoon.

Fertilisation and early development of the zygote take place in the female mosquito. She then feeds on another person and injects the parasite in her saliva. She acts as a **vector** for the parasite.

The immune system is unable to destroy the parasite because:
● it spends much time inside liver cells or red blood cells and so antigens on its surface are not exposed and cannot easily stimulate an immune response;
● the antigens on the surface of the different forms of the parasite are different.

Several sets of antibodies are therefore needed to combat malaria.

Schistosomiasis [AQA A Human Biology only]

Schistosomiasis or **bilharzia** is a parasitic disease caused by a group of flatworms of the genus *Schistosoma*. The adult worms (both female and male) usually settle in veins of the bladder or intestine. After fertilisation the female releases eggs that pass out with the urine or faeces. Whilst in the blood vessels, *Schistosoma* survives by:
● being firmly attached by suckers that prevent the flow of blood from dislodging it;
● covering itself with molecules from red blood cells to avoid detection by the immune system.

1 (a) The graph shows the growth of a bacterial population grown in a liquid culture at 20°C.

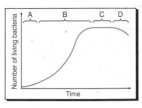

    (i) Name the phases labelled A and B. (2)

    (ii) How would the curve differ if the bacteria had been grown at 30°C? (2)

    (iii) Give two reasons for the decline in bacterial numbers. (2)

(b) The average number of bacteria in one of the small squares of a haemocytometer is 6. The volume of liquid in a small square is 0.00025 **mm³**. What is the number of bacteria per **cm³**? (2)

2 (a) List four ways in which bacteria can gain entry to the body. (4)

(b) What is the difference between endotoxins and exotoxins? (2)

3 (a) Label the diagram of the human immunodeficiency virus (HIV). (3)

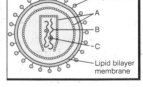

(b) Complete the following paragraph.

HIV injects RNA and .... into a helper T .... Inside this cell, the .... makes a DNA copy of the RNA. This becomes part of the host cell's DNA and is activated by DNA .... or by .... When activated, the HIV DNA directs the manufacture of HIV .... Some of these are used to make new ...., others become incorporated into the membrane of the helper T .... Here they stimulate an .... and the infected cell is destroyed by other .... (10)

4 Describe measures that can reduce the transmission of:

    (a) Salmonella; (3)   **(b)** TB. (2)

5 (a) What is the role of the female mosquito in the life cycle of the malarial parasite? (2)

(b) Explain two ways in which the malarial parasite is adapted to survive in humans. (4)

The answers are on page 116.

[AQA A Human Biology only]

Heart disease

A **heart attack** occurs when a region of **cardiac muscle** (making up the heart wall) receives a significantly reduced supply of oxygen. This happens when one of the coronary arteries becomes narrowed or blocked. Under these conditions:

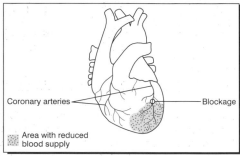

Coronary arteries

Blockage

Area with reduced blood supply

- the region with the reduced blood supply receives little oxygen;
- this region cannot respire aerobically;
- it respires anaerobically and produces **lactate** (lactic acid);
- the lactate and reduced energy supply cause muscle fatigue and the heart muscle contracts irregularly;
- some muscle tissue may die, this is known as **myocardial infarction**;
- the heart beat becomes weak and unco-ordinated, if a large area of the heart is involved it may stop beating altogether.

Atherosclerosis

Atherosclerosis is a progressive build up of fatty substances in the endothelium of an artery. Atherosclerosis in a coronary artery is a major risk factor in causing a heart attack.

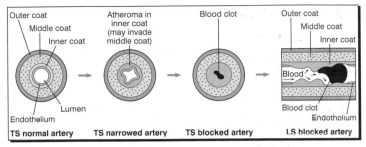

Outer coat	Atheroma in inner coat (may invade middle coat)	Blood clot	Outer coat
TS normal artery	TS narrowed artery	TS blocked artery	LS blocked artery

The regions where the fatty deposits build up are called **atheromas** or **plaques**. A blood clot (a **thrombus**) may also form in one of these regions because the plaque/atheroma stimulates platelets to release substances that promote blood clotting. The condition is called **thrombosis**: if it occurs in a coronary artery, it is called **coronary thrombosis**.

Thrombosis causes an increase in blood pressure and also weakens the wall of the artery. The increased pressure on the weakened wall causes ballooning of the artery. This is an **aneurysm**. If it bursts, it can be fatal.

Factors that increase the risk of atherosclerosis increase the risk of heart disease.

Hypertension

Hypertension means high blood pressure sustained over a considerable period of time. It damages the endothelium (lining) of arteries in particular and increases the risk of atherosclerosis. The sustained high pressure results in the artery wall developing a thicker muscle layer to resist the pressure. The smaller lumen increases resistance to blood flow and the pressure becomes even higher.

Smoking

Smoking raises blood pressure and so causes, indirectly, the effects already described. It also increases the levels of **fibrinogen** (one of the proteins involved in blood clotting) in the plasma and so increases the risk of a blood clot forming. Smoking increases levels of cholesterol in the plasma.

Level of plasma cholesterol

Cholesterol is one of the main components of the atheromas that form during atherosclerosis. High levels of plasma cholesterol are associated with higher levels of atherosclerosis, leading to increased risk of blood clotting as the damaged inner walls of arteries stimulate platelets to release substances that promote blood clotting.

The three factors listed all influence the risk of atherosclerosis. Your lifestyle can be adapted to minimise their effects.

- Eat a balanced diet low in animal fats (to reduce plasma cholesterol levels).
- Take regular exercise (to reduce blood pressure).
- Don't smoke.

Cancer

Cancer comprises a group of disorders that result from uncontrolled division of cells. Such division forms a cell mass called a **tumour**. Tumours can be **benign** or **malignant**.

Benign tumours are slow growing and are often contained within a fibrous capsule and so do not invade the tissue in which they originate. They never spread but remain in the area in which they originated. Although they do not invade surrounding tissue, a large benign tumour may cause damage by exerting pressure on blood vessels and restricting blood flow to the area. Benign tumours are not cancerous.

Malignant tumours are usually quicker growing and do invade the tissue in which they develop. They develop their own blood supply and compete with surrounding tissue for nutrients and oxygen. In addition, they often break out of this area and spread in the blood or the lymph to initiate secondary cancers in other organs. This is called **metastasis**. Malignant tumours are cancers.

All tumours arise from uncontrolled cell division. In a normal cell, mutations in two sorts of genes are involved in producing tumours.

Oncogenes are a group of genes that code for proteins that 'switch on' cell division. Normally the protein is inactivated and cell division remains switched off. Some mutated oncogenes produce a protein that cannot be inactivated but still switches on cell division.

Tumour suppressor genes are a group of genes that slow down the division of cells in a developing tumour. Mutation of these genes in tumour cells means that the tumour is more likely to develop into a significant growth.

Many tumours are detected and destroyed by the immune system. However, if the cells are too similar to ordinary body cells, the immune system may not detect them until the tumour is well established.

The rate of mutation of these genes is increased by the following.
- **Ionising radiation**, such as ultra-violet (UV) radiation and X-rays. Sunbathing in strong sunlight exposes the skin to UV and increases the risk of skin cancers such as malignant melanoma, which is often fatal. The ionising radiation in nuclear fall-out can penetrate all body tissues, not just the skin, so can cause leukaemias, etc.
- **Chemical carcinogens**, such as some of the chemicals found in cigarette smoke and in diesel exhaust fumes, increase the rate of mutation of genes in cells that are exposed to them. Smoking significantly increases the risk of lung cancer (the carcinogens are in the tar produced, not the nicotine).
- **Viral infections**; some viruses have been linked to particular cancers, e.g. some viruses causing genital warts are linked to cancer of the cervix.

1 (a) Explain how a blockage in a coronary artery can lead to a heart attack. (4)

(b) What is meant by:

   (i) atherosclerosis; (3)

   (ii) coronary thrombosis? (3)

2 Complete the following paragraph.

Smoking is one of the .... associated with coronary heart disease. The main effects of smoking are to raise .... and to increase levels of .... and .... Hypertension is a condition in which blood pressure is continually .... This results in damage to the .... of arteries which increases the risk of .... Because of the increased .... the .... layer in the wall of the artery becomes .... and the lumen .... This raises the .... still further. High levels of plasma cholesterol increase the rate of .... and the risk of .... forming. (14)

3 (a) Give three differences between a benign and a malignant tumour. (3)

(b) Explain how mutations in oncogenes and tumour suppressor genes allow tumours to form. (4)

(c) Give three factors that increase the rate of mutation of oncogenes and tumour suppressor genes. (3)

The answers are on page 117.

Socio-economic measures

Look at the graphs. **Medical intervention** played an important role in the virtual elimination of **diphtheria**, but the decline in the incidence of **tuberculosis (TB)** was well underway before the vaccination programme began.

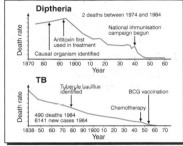

The reduction in deaths from TB, and other diseases such as whooping cough and measles, is probably due to factors such as:

- improved diet giving increased resistance to disease;
- improved living conditions with less overcrowding and so less opportunity for the spread of disease;
- improved sanitation and personal hygiene reducing the opportunity for transmission of disease;
- better education, making people aware of the nature of disease and how to reduce transmission, e.g. knowledge that sexual transmission of some diseases can be reduced by using condoms.

Recently there has been an increase in the incidence of TB. This is, in part, due to the increased transmission to AIDS sufferers. The World Health Organisation estimates that 1.5 million cases of TB each year are due to the reduced immunity of AIDS sufferers.

Medical intervention

Vaccination

Most **vaccines** consist of a preparation of one of the following:
- an **attenuated** (weakened) strain of the pathogen, e.g. for polio and measles;
- dead **pathogens**, e.g. for whooping cough and typhoid;
- just the **antigens** found on the micro-organism, e.g. for influenza.

Vaccination stimulates an immune response (see page 91).

CONTROLLING DISEASE (2)

The aim of vaccination is to reduce the spread of disease by increasing the numbers of immune individuals in a population. Once this exceeds 90%, transmission of the disease becomes very difficult. Mass vaccination programmes have succeeded in eliminating smallpox from the planet.

Antibiotics

Antibiotics, such as penicillin, are used to treat a range of bacterial infections. They act in one of the following ways.

Mode of action	Example	Effect
Interferes with cell wall synthesis	Penicillin	Weakens cell wall, resulting in osmotic lysis (water enters by osmosis and pressure bursts cell)
Prevents DNA replication	Nalidixic acid	Bacteria are not killed but are unable to multiply
Prevents mRNA synthesis	Rifamycin	Kills bacteria as no enzymes made to control metabolic reactions
Prevents transfer of amino acids to ribosomes	Tetracycline	Kills bacteria as no enzymes made to control metabolic reactions

Because viruses have no cell walls or organelles, antibiotics are ineffective against them.

Beta-blockers and heart disease

Beta-blockers are a group of drugs used to treat angina and hypertension (see page 97).

The hormone adrenaline acts by binding to beta receptors in the SA node and in the muscle in the ventricle walls. When it binds, it causes the SA node to increase the heart rate and the ventricle walls to contract with more force.

Beta-blockers have a shape that can bind to these receptors and so prevent adrenaline from binding. They therefore reduce blood pressure and prevent the heart rate from rising to dangerous levels.

1 Name and explain three socio-economic factors that have helped to reduce the incidence of disease. (6)

2 **(a)** What is the aim of a vaccination programme? (2)
(b) Explain why a vaccination makes a person immune to the particular disease vaccinated against, but only that disease. (4)

3 **(a)** Bacteriostatic antibiotics stop bacteria from reproducing, bactericidal antibiotics kill bacteria. Name one antibiotic of each type, describe its mode of action and the effect of its action. (6)
(b) Explain why antibiotics are ineffective against viruses. (2)

4 Explain how beta blockers reduce hypertension. (4)

The answers are on page 118.

Check yourself answers

1 (a) Higher resolution (1), due to shorter wavelength of electrons (1). Remember that higher magnification does not produce more detail in itself. You can enlarge a light micrograph many times, but you will not see any more detail.

   (b) Any two of: specimen not alive; appearance may be unnatural; can only see thin section.

2 Apparent size = 6.5 cm = 65 mm = 65 000 μm (1); magnification = apparent / real (1) = 65 000 / 130 = 500 (1) Make sure that both measurements are in the same units.

3 (a) A, nucleus (1); B, Golgi apparatus (1); C, endoplasmic reticulum (rough) (1).

   (b) Many ribosomes (synthesise proteins) (1); many vesicles (export proteins) (1). You need to know about the different organelles involved with protein synthesis and transport within a cell: they will all be abundant in a cell that makes a lot of protein to be 'exported'.

4 (a) Keep pH constant (1); changes would affect enzyme/protein structure (1).

   (b) Buffer is ice-cold to minimise enzyme reactions (1), which might damage organelles (1), and is isotonic to prevent osmotic water uptake/loss by organelles (1), which would burst or shrink the organelles (1).
   It is the organelles (not the whole cell) that might be damaged osmotically if the buffer was not isotonic with (same concentration as) the contents of the cell. The cells are being homogenised – they will be in shreds anyway!

5 (a) Tissue: many similar cells carrying out the same function (1). Organ: several tissues combined in a single structure with an identifiable function, the tissues all contribute to that function (1).

   (b) Artery is made from several tissues (smooth muscle, fibrous tissue, elastic tissue and endothelium) (1); capillary only has endothelium (1).

   (c) Most contain just one type of cell (1); blood contains several types of cell (1).

6 Any six of: the leaf is broad and flat, providing a large surface area to absorb light; it is thin, so there is a small diffusion distances for gases; it is easy for light to penetrate; many chloroplasts allow efficient light absorption; chloroplasts are mainly near upper surface for efficient absorption; spaces in spongy mesophyll allow gas exchange; stomata control gas exchange; contains xylem to bring water. Look at the mark allocation: in a question like this, 6 marks = 6 features.

BIOLOGICAL MOLECULES (page 11)

1 (a) Carbohydrates (1) sugars (1); monosaccharides have one ring (1); disaccharides have two rings (1).

   (b) Glycogen is more branched than starch, and more branched than cellulose (1); glycogen is made from α-glucose (1), cellulose from β-glucose (1).

   (c) Glycosidic (1).

Check yourself answers

2 **(a)** Compact: it can store a lot in a small place (1); insoluble: it has no osmotic effects/does not move from storage cells (1).

(b) Add iodine solution (1); would turn blue/black (1).

(c) Parallel molecules with hydrogen bonds (1); form fibrils/fibres (1); form a mesh (1).

3 **(a)** You can work out from where there are fewer bonds to hydrogen atoms where the double bond should be. You should know the COOH group.

(b) They form saturated fats (1), which are implicated in atherosclerosis/heart disease (1).

(c) **(i)** 3 (1);   **(ii)** 2 (1).

4 **(a)** Did you remember to include the water molecule?

(b) Condensation (1). Think of two structures being condensed into one.

(c) Reducing: any one from glucose/fructose/maltose/lactose (1); non-reducing: sucrose (1).

5 **(a)** Spot must be above the solvent level (1), otherwise it will dissolve away (1); do not let solvent run to end of paper (1), spreading will be inaccurate (1).

(b) $x = \dfrac{1.25}{1.5} = 0.8$ (1). $y = \dfrac{0.75}{1.5} = 0.5$ (1).

An R_f value is always a fraction. If your answer is greater than 1, you got the figures upside down!

6 Any three examples and explanations from the table on page 10.

1 Lower the activation energy (1); more molecules have enough energy to react (1). Many candidates only give the first point, which is only half the answer.

2 **(a)** Enzymes only catalyse one reaction/only bind with one substrate (1).

(b) The active site has a specific shape (1); complementary to that of the substrate (1); only the substrate can fit/bind with the active site (1).

(c) In induced fit, the active site does not initially 'fit' the substrate (1); it changes shape to fit (1).

The principle of enzyme action is not difficult: a substrate molecule fits inside part of the enzyme (active site). This part must either be already shaped to accommodate the substrate (lock and key) or change shape to accommodate it (induced fit).

3 (a) (i) Increasing temperature means more kinetic energy (1); molecules move faster (1); they collide more often (1).

(ii) Optimum temperature (1); fastest reaction rate (1).

(iii) Denatured (1); too much energy (1); breaks bonds/alters shape of active site (1).

(b) (i) Increasing concentration increases rate (1); as some active sites not occupied all the time (1); more substrate means more active sites occupied/reacting (1).

(ii) All active sites are occupied all the time (1); extra substrate cannot increase the number of active sites occupied/reacting (1).

There is quite a lot of detail to be aware of in answering these sorts of questions. You need to learn to recognise the graphs and be aware of what each represents and why.

4 The missing words are: active site (1); complementary (1); enzyme–substrate complex (1); competitive (1); active site (1); more/greater/increased (1); allosteric site (1); shape (1); active site (1); non-competitive (1); independent (1). This paragraph forms a useful revision of inhibition.

5 Competitive (1); inhibition changes with substrate concentration (1); non-competitive is independent of substrate concentration (1). This is a fairly standard examination question: learn to recognise the graph and be able to account for its shape.

6 (a) B, C and D (1).

(b) Cells of the micro-organism must be burst open (1); more complex separation techniques are needed (1).

(c) Easier to optimise conditions (1); less costly downstream processing (1).

7 (a) Any three of: continuous production is possible (1); the enzyme does not contaminate the product (1); the enzyme can be reused many times (1); the enzyme is more stable, so higher temperatures are possible (1).

(b) Use in production of specific products such as lactose reduced milk and fructose syrup (1); use in biosensors such as clinistix strips (1).

THE CELL CYCLE (page 23)

1 A, DNA (1); B, centromere (1); C, histone (1); D, two chromatids (1). The double strand in the centre identifies the DNA, this has histone molecules bound to it. The two chromatids making up one chromosome are joined by the centromere.

2 Pairs of chromosomes (1); with the same genes (1).

3 Chromosomes in homologous pairs originate from different parents (1); chromatids are formed by duplication of the original chromosome (1); DNA replicates (1); therefore contain identical genetic material (1). DNA is the genetic material, and when it replicates the two new molecules are identical (see replication of DNA on page 29).

Check yourself answers

4 Ribosomes (1); to synthesise proteins (1); mitochondria (1); to provide necessary ATP (energy) (1). You should know that protein synthesis is taking place in G_1, that it takes place in ribosomes and that it is an anabolic process needing ATP from respiration in mitochondria.

5 (a) A – C – D – B – F – E (1). A is the first (chromosomes double in pairs, one nucleus in one cell). E is the last, two cells each with a nucleus. If you are really not sure, work forwards from A (C still has the chromosomes double and joined) and backwards from E (F has two nuclei but not yet two cells) until you have the sequence.
  (b) Spindle (1).

6 Any three of:

Feature	Mitosis	Meiosis
Number of divisions	One	Two
Number of cells formed	Two	Four
Diploid/haploid daughter cells	Diploid	Haploid
Variation in daughter cells?	No – are genetically identical	Yes

7 Meiosis ensures haploid gametes (1). Fertilisation restores diploid condition (1). Mitosis 'copies' zygote/maintains diploid condition (1).

MEMBRANES AND TRANSPORT (page 28)

1 (a) A, phospholipid bilayer (1); B, carrier protein (1); C, ion channel (1); D, polysaccharide chain (1); E, cholesterol (1).
  (b) Na^+, ion channel (C) (1); glucose, carrier protein (B) (1); glycerol, phospholipid bilayer (A) (1).
  A glucose molecule is large/water soluble; glycerol is soluble in lipids.
  (c) Denature proteins (1); alter structure of membrane (1); 'gaps' appear (1).
2 (a) Compartmentalisation (1); reaction surface / for enzyme to control reactions (1).
  (b) Any four of: around cell; around nucleus; around other organelle (mitochondrion/ chloroplast /lysosome); endoplasmic reticulum; Golgi apparatus.
  (c) Around nucleus (1); around organelles (1); endoplasmic reticulum (1).
3 Alveoli have large surface area (1); epithelium is thin (1); large concentration difference between air in alveolus and blood (1).
4 (a) A, facilitated diffusion (1); carrier protein does not use ATP (1). B, active transport (1); carrier protein uses ATP (1).
  (b) Active transport removes amino acids from cell (1); maintains a low concentration in cell (1).

Check yourself answers

5 **(a)** Arrows from: A → B (1); A → C (1); B → C (1).
Water always moves to a **more** negative water potential ψ.

(b) ψ of red blood cells lower than/more negative than distilled water (1);
water moves in by osmosis (1); red cells swell – membrane cannot resist
swelling (1).

(c) ψ of red blood cells higher than/less negative than strong salt solution
(1); water moves out by osmosis (1); cells shrink (1).

1 **(a)** A, cytosine (1); B, hydrogen bond (1); C, pentose sugar (deoxyribose) (1);
D, phosphate group (1); E, nucleotide (1); F, thymine (1). You must be able
to label a diagram of the DNA molecule.

(b) Strands where bases bind to their partner (1), e.g. T binds to A, C binds
to G (1), have equal amounts of T and A, C and G (1).

(c) **(i)** A = T, therefore A = 22% (1)
(ii) A + T + C + G = 100% (1), A + T = 44% therefore C + G = 56% (1),
C = G therefore C = 56 ÷ 2 = 28% (1)

This is a popular examination question, make sure you understand the maths.

2 **(a)** A, DNA (1); B, mRNA (1); C, protein/polypeptide (1); D, amino acid (1); E,
tRNA (1) D and E could be either way round.

(b) Transcription: nucleus (1); translation: ribosomes/rough ER (1). Again, you
need to be able to complete flow charts like this one.

3 **(a)** GGC TTG CCT TAT ATG (1) Just think of the complementary bases.
(b) First codon, CCG (1). Fourth codon, UAU (1). Again base pairing, but
remember that U replaces T in all RNA molecules.

4 The missing words are: S (1); semi-conservative (1); strand (1); helicase (1);
hydrogen (1); polymerase (1); complementary (1); nucleotides (1); base pairing (1).

5 A, mRNA (1); B, codon (1); C , anticodon (1); D, peptide bond (1).

There are a number of other styles of questions you could be asked in an examination,
but the ones above represent the basics that you must be able to answer.

1 **(a)** A, restriction (endonuclease) (1); B, ligase (1).
(b) So that they can combine/bind easily (1).
(c) Plasmids that have extra DNA (1); from another organism (1).
(d) They contain the resistance gene (1); so can grow even when the
antibiotic is present (1).

You may well be given a diagram very like this one to label or explain. Make
sure that you can recognise all the key stages in this flow chart.

Check yourself answers

2 The missing words are: enzyme (1); DNA (1); DNA polymerase (1); DNA nucleotides (1); complementary (1); hydrogen (1).

3 Can optimise conditions for each phase separately (1); different nutrients are needed for growth and for production (1); nutrients not wasted in each phase (1).

4 **(a)** A thermostable enzyme allows continuous production (1); as it does not denature at high temperatures (1); high optimum allows high reaction rates (1); primers allow faster replication (1); by locating starting sequences (1).

   (b) Makes many copies of a small sample of DNA (1); exactly like the original (1). You may also have to compare the PCR reaction with the DNA replication in a living cell. Bear in mind that in the cell, no primers are used, the temperature is very different and DNA helicase splits the strands, not heat.

   (c) 8×2^6 (1) = 512 (1)

5 **(a)** 7 times (1). If you cut a piece of string seven times, you will end up with eight pieces.

   (b) Separated due to (molecular) mass (1); lightest/smallest travel furthest (1).

6 **(a)** Cut out with restriction (endonuclease) (1); leaving sticky ends (1).

   (b) Virus can infect cells/enter cells (1); and carry 'healthy' gene with it (1).

   (c) Cells with healthy gene are continually lost (1); replaced by cell division (1); new cells formed still have cystic fibrosis gene (1); treatment must be repeated (1). You could also be asked about the disadvantages of using a virus, such as the risk of disease transmission or inducing a cancer.

1 Residual volume, inspiratory capacity and expiratory reserve (1).

2 Factors on the top line of the formula (surface area, concentration difference) increase diffusion rate when they increase (1); thickness decreases the rate when it increases (1); the thin walls of alveoli and capillaries reduce thickness of membrane (1); large numbers of alveoli and capillaries create a large exchange surface (1); ventilation and circulation maintain a high concentration difference (1). The key to answering this question successfully (it has appeared several times in examinations) is to relate the formula to the actual situation. Show how the various components of the formula correspond to real structures/processes.

3 The missing words are: medulla (1); impulses (1); inspiratory centre (1); inhibit (1); diaphragm (1); intercostal muscles (1); inspiratory centre (1); passively (1) stretch receptors (1); inhibition (1); inhalation (1).

4 **(a)** Has higher affinity for O_2 than adult Hb (1); adult Hb can be only 20% saturated, so unloads O_2 (1); fetal Hb can be 80% saturated and so loads free O_2 (1).

(b) Needed to make haemoglobin (Hb) molecule (1).If Hb in the lungs is 95% saturated, 95% is carrying O_2. If it is only 20% saturated in the tissues, 75% (95 – 20) must have unloaded O_2.

5 Light triggers K^+ pump and ions move into guard cells (1); water follows (1); by osmosis (1); guard cells become turgid/swell (1); and open the stoma (1); in dark, K^+ ions move out (1); water follows (1); guard cells shrink and close stoma (1).

TRANSPORT IN HUMANS   (page 50)

1 Whale is much larger (1); has smaller surface area/volume ratio (1); cannot obtain oxygen through surface/has lungs (1); oxygen must be transported from lungs (1).

2 Arteries have more muscle (1); to withstand higher pressure (1); arteries have more elastic tissue (1); to recoil from being stretched (1); veins have larger lumen (1); to reduce resistance to blood flow (1); veins have valves (1); to prevent backflow of blood (1).

3 (a) You may have to work out some information, e.g. blood pressure, from a graph and then relate this to other features, such as opening/closing of valves.

	Atrio-ventricular valve (open/closed)	Aortic valve (open/closed)	Blood pressure in ventricle (high/low)
Peak of atrial systole	Open (1)	Closed (1)	Low (1)
Peak ventricular systole	Closed (1)	Open (1)	High (1)

(b) Stroke volume increase (1); heart rate increases (1); due to sympathetic impulses (1); and hormone adrenaline (1).

(c) Cardiac output increased (1); arterioles to brain constrict (1); same volume is a smaller fraction/proportion (1). You must be able to distinguish between amount and proportion.

4 The missing words are: myogenic (1); S–A (1); Purkyne tissue (1); contract (1); ventricles (1); A–V (1); slowly (1); bundles of His (1); contraction (1); A–V (1); atria (1); ventricles (1); sympathetic (1); cardiac centre (1); adrenaline (1); parasympathetic (1).

5 (a) Foreign antigens recognised (1); B cells selected (1); B cells cloned (1); some become plasma cells (1); secrete antibodies (1); some become memory cells (1).

(b) Active involves immune response (1); antibodies manufactured (1); in passive immunity, antibodies are acquired (1).

6 At arterial end, hydrostatic pressure higher in capillary (1); Ψ less negative in tissue fluid (1); hydrostatic pressure greater (1); fluid forced out (1); at venous end Ψ of tissue fluid less negative (1); greater effect than hydrostatic pressure (1); H_2O returns (1).

1 **(a)** Both made from tubular cells (1); arranged end to end (1).
   (b) Xylem cells are dead (1); empty (1); end wall not present (1).

2 **(a)** A, apoplast (1); B, symplast/vacuolar (1). Some texts differentiate between the pathway through the cytoplasm only (symplast) and that through cytoplasm and vacuoles (vacuolar).
   (b) Cell walls (1); of cells in endodermis (1); act as apoplast block (1).
   (c) ψ soil is less negative than ψ root epidermis/root hair cells (1); ψ root hair cells less negative than ψ xylem cells (1); water moves down ψ gradient (1).

3 The missing words are: stomata (1); vapour (1); air spaces (1); evaporation (1); reduces (1); osmosis (1); tension (1); upwards (1); cohesion (1); cohesion–tension (1); transpiration (1).

4 Humidity (1); humid atmosphere reduces diffusion gradient, slows transpiration (1); temperature (1); high temperature increases kinetic energy of water vapour molecules, move away from stomata, increasing diffusion gradient and transpiration rate (1); light intensity (1); bright light stimulates opening of stomata, increasing transpiration rate (1); air movement (1); wind blows water vapour molecules away from the stomata, increasing the diffusion gradient and transpiration rate (1). Did you explain the effects of each factor on the diffusion gradient (difference in concentrations of water vapour inside and outside the leaf) where relevant?

5 Reduced leaves (1); fewer stomata to lose water vapour (1); surface hairs (1); trap water vapour and reduce diffusion gradient (1); curled leaves (1); trap water vapour inside 'tube' and reduce diffusion gradient (1); sunken stomata (1); not exposed to high temperatures/winds which would increase diffusion gradient and transpiration (1); restricted stomatal opening times (1); water can only be lost when open (1). Did you give only adaptations that reduce water loss as the question asked, or did you include features like storing water in swollen stems?

6 **(a)** Sugars from leaves (1); translocated downwards in phloem (1); no phloem in ringed area (1); cannot pass (1); sugars accumulate and stem swells (1).
   (b) Sucrose pumped into phloem in leaf (1); water follows (1); creates hydrostatic pressure (1); drives liquid to sink (1); with low hydrostatic pressure (1); sucrose unloaded (1).
   (c) Supply C^{14} in CO_2 (1); sample at regular intervals (1); check for radioactivity in phloem (1); with Geiger counter (1).

Check yourself answers

1 **(a)** Large molecules to smaller ones (1); by hydrolysis (1); smaller ones are soluble (1).

   (b) Large molecules cannot cross gut wall (1); into blood stream (1).

2

Region of gut	Secretion	Enzyme(s) & other contents	Digestive action
Buccal cavity	Saliva	Amylase	*Starch → maltose* (1)
Stomach	*Gastric juice* (1)	*Pepsin* (1)	Protein → short chain peptides
		Hydrochloric acid	*Kills micro-organisms, provides, optimum pH for pepsin* (1)
Lumen of small intestine	Bile	Bile salts	*Emulsify lipids* (1)
		Sodium hydrogen-carbonate (1)	Provides alkaline pH
	Pancreatic juice	Lipase	*Triglycerides → fatty acids & glycerol* (1)
		Amylase (1)	Starch → maltose
		Trypsin	Proteins → short chain peptides
		Exopeptidases	Short chain peptides → dipeptides
Cells in wall of ileum		Maltase	*Maltose → glucose* (1)
		Dipeptidase (1)	Dipeptides → amino acids

3 **(a)** Protein digesting enzymes (1); cells of stomach/pancreas contain protein (1); would digest/hydrolyse cells (1).

   (b) Endopeptidases digest peptide bonds in middle of molecule (1); create 'more ends' (1); larger surface area for exopeptidases to act on (1).

4 **(a)** Villi and microvilli give large surface area (1); intestinal wall is thin giving short diffusion distance (1); good blood supply for maximum absorption (1).

   (b) Removal of glucose to blood depends on active transport (1); and facilitated diffusion (1); maintains a low concentration in cell (1); more can enter from gut by facilitated diffusion (1).

5 **(a)** Conditioned reflex ensures that saliva is released before food is ingested (1); simple reflex ensures continued secretion while food is present in mouth (1).

   (b) Released from wall of duodenum (1); in response to acid food from stomach (1); stimulates release of pancreatic juice (1).

Check yourself answers

1   **(a)**   Autotrophic nutrition involves synthesis of organic molecules from inorganic ones (1); heterotrophic nutrition involves intake of organic molecules (1).

    (b)   Autotrophic feeders: plants (photo-autotrophs) (1); nitrifying bacteria (chemo-autotrophs) (1). Heterotrophic feeders: any two of animals (1); saprobonts/saprophytes/decomposers (1); parasites (1).

2   The missing words are: symbiosis (1); host (1); endoparasites (1); host (1); ectoparasites (1); food (1); host (1); decay (1); dead (1); enzymes (1); digestion (1).

3   **(a)**   Nutrition in which organic materials are ingested/eaten (1); into a gut for digestion (1).

    (b)   Herbivore molars are ridged, carnivores are pointed (1); herbivore incisors are chisel-like, carnivores are pointed (1); carnivores have large canines, these are often absent in herbivores (1).

    (c)   Jaw action of herbivore is side to side (1); action of carnivore is up and down (1).

1   **(a)**   Self-contained system (1); in which living things interact with each other (1); and with their physical environment (1).

    (b)   **(i)** Population is all the individuals of one species in an ecosystem at a given moment (1); community is all the individuals of all species in an ecosystem at a given moment (1).
    (ii) Habitat is the area occupied by an organism (1); niche is the role fulfilled in a habitat by an organism (1).
    (iii) Biotic factors are related to living things, abiotic to non-living things (1).

2   **(a)**   Producer – plankton (1); secondary consumer – fish (1).

    (b)   Correct shape (1); each level labelled correctly (1).

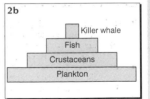

3   The missing words are: photosynthesis (1); reflected (1); wavelength (1); organic molecules (1); photosynthesis (1); trophic (1); decomposers (1); 10 (1); food chain (1); energy (1).

4   A – photosynthesis (1); B – excretion (1); C – decay (1); D - respiration (1).

5   **(a)**   **(i)** In summer there are longer days/more sunlight (1); there are more leaves on trees (1); there is more photosynthesis than respiration (1); more CO_2 is absorbed than released (1); levels in the atmosphere fall (1). In winter all the processes are reversed and you could equally well have scored full marks by describing what happens in winter.

(ii) Deforestation (1) reduces plants available to absorb CO_2 (1); combustion of fossil fuels (1) adds CO_2 to atmosphere (1).

(b) Any three of: increased sea levels (1), due to polar ice caps melting (1); long-term climate change (1), due to changed rainfall and temperature (1); changes in ecosystems (1), better adapted organisms enter and outcompete existing ones (1); extinctions (1), some species are out-competed and unable to colonise new areas (1).

6 (a) **(i)** Nitrates used for protein synthesis (1); more nitrates more proteins (1); increased reproduction of algae (1).

(ii) Algae die when nitrates are used up (1); decomposers decay algae and multiply (1).

(iii) Increased respiration (1); of decomposers (1); uses much oxygen (1).

(b) Decomposers multiply (1); respire more (1); use much oxygen (1).

7 (a) Random quadrats of known area (1); find mean number per quadrat (1); find area of site under investigation (1); multiply mean by ratio of the two areas (1).

(b) Capture a sample of the animals, count them (N_1) mark them and release (1); allow time to mix and recapture a second sample (1); count them (N_2) and the number marked (n) (1); population is $\dfrac{N_1 \times N_2}{n}$ (1).

8 (a) Any three of: sunken stomata (1) reduce transpiration (1); C_4 photosynthesis (1) is more efficient at high temperatures (1); extensive root systems (1) obtain much more water when it rains (1); leaves reduced to spines (1) reduce transpiration (1), water storage in stems (1) allows survival for longer periods (1), curled leaves (1) increase humidity inside the rolled leaf which reduces transpiration (1).

(b) **(i)** Any two of: temperature tolerance (1), reduced water loss in sweat (1); small amounts of concentrated urine (1), reduced water loss in urine (1); large ears (1), large area of effective heat loss (1).

(ii) Any two of: thick fur for increased insulation (1); thick layer subcutaneous fat for increased insulation (1); large bodies so small SA/V ratio reduces heat loss (1).

SEXUAL REPRODUCTION IN FLOWERING PLANTS (page 78)

1 (a) A – sepal (1); B – anther (1); C – petal (1); D – stigma (1); E – style (1); F – ovary (1).

(b) **(i)** C (1); **(ii)** F (1).

2 (a) Transfer of pollen (1); from anther to stigma (1).

(b) Insect-pollinated flowers have: larger petals (1) to attract insects (1); nectaries (1) provide a 'reward' so insects will visit others of same type (1); stigmas placed inside flower (1) so insect must touch it and deposit pollen (1); sticky or hooked pollen (1) so that pollen attaches to the

REPRODUCTION IN FLOWERING PLANTS (page 78)

insect (1). Any three of the above will get you full marks. You could also have described the features of wind-pollinated plants in a comparative way.

3 The missing words are: pollen grain (1); tube (1); ovule (1); pollen grain (1); tube (1); micropyle (1); gametes (1); ovule (1); polar (1); zygote (1).

4 Separate male and female flowers (1); self-fertilisation is impossible (1). Stamens and ovaries mature at different times (1); when pollen tube reaches ovule of same flower, ovule has not developed or has already been fertilised (1). Self-incompatibility (1); if pollen tubes will not grow or grow only very slowly, male nuclei cannot reach ovule of same flower (1).

5 (a) Seed is a fertilised ovule (1); fruit is fertilised ovule plus ovary wall (1).

   (b) Any three of: wind (1); birds (1); other animals (1); water (1); explosive mechanisms (1).

REPRODUCTION IN HUMANS AND OTHER MAMMALS (page 84)

1 Similarities, any two of: both involve mitosis and meiosis (1); both have a maturation phase (1); both produce haploid gametes (1); both have a growth phase (1). Differences, any three of: multiplication and growth phases take place before birth in females (1); from puberty, sperm production is continuous, oocyte maturation is intermittent (1); one ovum from a primary oocyte, four sperm from a primary spermatocyte (1); gametogenesis produces millions of sperm but only thousands of ova (1).

2 The missing words are: ovum (1); oocyte (1); ovum (1); meiotic (1); hydrolytic (1); glycoprotein (1); zona pellucida (1); passage (1); membrane (1); oocyte (1); meiotic (1); oocyte (1); ovum (1); ovum (1); zygote (1). Make sure you can distinguish between an **oocyte** and an **ovum**.

3 (a) Hollow (1); ball of cells (1); with inner cell mass at one end (1).

   (b) Folded membrane (1); gives large surface area for exchange (1); thin membrane (1); gives short diffusion distance (1); fetal circulation in placenta (1); maintains a concentration gradient (1).

4 (a) A – FSH (1), levels rise before any other hormone (1); B – LH (1), levels rise and peak on day 14 (1); C – oestrogen (1), levels rise after FSH (1); D – progesterone (1), levels rise only after ovulation (1). You must be able to identify the hormones from graphs like this. Familiarise yourself with the reasons for identifying them.

   (b) Ovulation (1).

   (c) Menstruation (1).

5 (a) Reduction in levels of progesterone (1); which inhibits muscle contraction (1).

   (b) Reduction in levels of oestrogen and progesterone (1); maintain lining of uterus (1); lack of these leads to breakdown of uterus lining (1).

6 (a) Fertilisation outside body (1); using females own egg (1); treat female infertility (1).

Check yourself answers

(b) All sheep given injections of progesterone (1); oestrous cycle is 'suspended' (1); as it inhibits secretion of FSH and LH (1); when injections stop, new cycle starts in all sheep (1); at the same time (1); farmer can arrange for insemination of all flock at same time/all lambs produced at same time (1).

HUMAN DEVELOPMENT AND AGEING   (page 88)

1 (a) Relative growth rate is the increase in body mass/height per unit time (1); absolute growth is the total mass/height at any one time (1).
   (b) Changes appear at puberty (1); females go though puberty earlier than males (1); secretion of sex hormones causes increased secretion of growth hormone (1); males finish growth spurt later (1); have bigger bodies than females (1).
2 (a) A (1); childhood level is higher than adult level (1).
   (b) B (1); reaches nearly full size by age 8–10 (1).
3 (a) The missing words are: sperm (1); 70 (1); 45–50 (1); menopause (1); FSH (1); oocytes (1); oestrogen (1); progesterone (1); menstruation (1); HRT (1); menopause (1); oestrogen (1).
   (b) Any four of: decreased BMR (1); decreased nerve conduction velocity (1); decreased cardiac output (1); decreased respiratory capacity (1); osteoporosis (1); osteoarthritis (1); decreased fertility (1).
4 (a) Replaces oestrogen loss resulting from menopause (1); encourages cell division (1); of bone cells (1).
   (b) Cartilage is lost from the articulating surfaces of the joint (1); numbers of bone-forming cells increases (1); more bone is produced (1); movement is more restricted and more painful (1).

PATHOGENS AND DISEASE   (page 95)

1 (a) (i) A – lag phase (1); B – log phase (1).
        (ii) Rise more steeply/increase more quickly (1); to same maximum level (1).
        (iii) Decline in nutrients/oxygen (1); accumulation of toxic excretory products (1).
   (b) Number in 1 mm^3 = 6/0.00025 = 24,000 (1). There are 1000 mm^3 in 1 cm^3 so number in 1 cm^3 = 1000 × 24,000 = 24,000,000 (1). With most calculations there is 1 mark for the correct answer and 1 for the working out.
2 (a) Damaged skin (1); being breathed in (1); contaminated food/water (1); sexual intercourse (1).
   (b) Endotoxins – released when bacterium dies (1). Exotoxins – secreted during growth and reproduction of bacterium (1).

3 **(a)** A, protein coat (1); B, nucleic acid/RNA (1); C, reverse transcriptase (1).

(b) The missing words are: reverse transcriptase (1); lymphocyte (1); reverse transcriptase (1); replication (1); protein synthesis (1); proteins (1); viruses (1); lymphocyte (1); immune response (1); lymphocytes (1).

4 **(a)** Good personal hygiene (1); good institutional hygiene (1); cooking food thoroughly (1).

(b) Pasteurisation of milk (1); slaughter of infected cattle (1). Did you give measures that will prevent transmission rather than treat the illness?

5 **(a)** Vector/transmission from host to host (1); allows sexual reproduction of parasite/allows fertilisation to take place (1).

(b) Spends much time inside liver cells and red blood cells (1) not exposed to the immune system (1); different forms of the parasite have different surface antigens (1), immune system must make different antibodies to destroy them (1). Did you explain as well as describe the adaptations?

THE BIOLOGICAL BASIS OF HEART DISEASE AND CANCER (page 100)

1 **(a)** Restricts blood flow (1); region of heart muscle is deprived of oxygen (1); cannot respire aerobically (1); muscle stops contracting/muscle dies (1).

(b) **(i)** build up of fatty substances (1); in the endothelium/lining (1); of an artery (1).

(ii) blockage of an artery (1); in the wall of the heart (1); by a blood clot (1).

2 The missing words are: risk factors (1); blood pressure (1); plasma cholesterol (1); plasma fibrinogen (1); high/raised (1); endothelium/lining (1); atherosclerosis (1); pressure (1); muscle (1); thicker (1); narrower/smaller (1); blood pressure (1); atherosclerosis (1); coronary thrombosis (1).

3 **(a)** Benign tumours are: slower growing (1); encased in a capsule/do not invade tissues (1); do not spread to other areas of the body (1). You could make the same points by describing the opposite for malignant tumours. You don't usually have to do both.

(b) Mutated oncogenes produce a protein that switches on cell division (1); the protein cannot be inactivated (1); tumour begins to form (1); mutated tumour suppressor genes cannot slow down the growth of the tumour (1).

(c) Exposure to high energy radiation/UV light/X-rays/gamma rays (1); exposure to chemical carcinogens e.g. in tar from cigarette smoking (1); exposure to some viruses (1).

1 Improved diet (1); gives increased resistance to disease (1); improved living conditions/less overcrowding (1); less risk of transmission (1); improved standards of hygiene/sanitation (1); reduces reproduction of micro-organisms (1).

2 **(a)** Reduce the spread of a disease (1); by increasing the numbers of individuals who are immune (1).

(b) Antigens on the surface of the micro-organism (1); are unique/different to antigens on other micro-organisms (1); stimulate only one type of B lymphocyte (1); only one type of memory cell made (1). The question asks why we become immune to only that particular disease not how we destroy the invading micro-organism. In this question, references to antibodies are irrelevant.

3 **(a)** Bactericidal antibodies – penicillin (1) affects cell wall synthesis (1) resulting in osmotic lysis (1); or rifamycin (1) halts mRNA synthesis (1) so no enzymes control metabolism (1); or tetracycline (1) prevents tRNA taking amino acids to ribosomes (1) so no enzymes control metabolism (1). Bacteriostatic antibodies – nalidixic (1) prevents DNA replication (1) so bacteria cannot multiply (1).

(b) Viruses have no organelles/no ribosomes/no cell walls (1); viruses are found inside living cells so cannot be targeted (1).

4 Bind to receptors in heart muscle and SA node (1); prevent adrenaline binding (1); reduce rate of heart beat (1); reduce force of contractions (1).

INDEX

INDEX